Additional Praise for *In Light-Years There's No Hurry*

"In her own quest for perspective, Marjolijn van Heemstra beautifully weaves together strands of science, poetry, natural history, and philosophy to construct a possible lifeline for the disoriented and perplexed. A warm and elegant book."
—Meg Howrey, author of *They're Going to Love You* and *The Wanderers*

"Marjolijn van Heemstra offers a delightful poet's-eye view of the cosmos as she travels by train and through cyberspace to bring the cosmic perspective from space into modern reach. Her timely book urges us to seek solace in the darkness and turn our gaze upward to discover a shared sense of unity."
—Jacob Haqq-Misra, author of *Sovereign Mars: Transforming Our Values through Space Settlement*

"A rich and thoroughly enjoyable read. . . . Understanding the universe involves a certain cold scientific detachment, but Marjolijn van Heemstra takes us on a delightful and entertaining journey in which the inescapable human story and interpretation of what it all means is blended seamlessly with discovery."
—Charles S. Cockell, author of *Taxi from Another Planet* and *The Equations of Life*

IN

LIGHT-YEARS

THERE'S

NO HURRY

ALSO BY MARJOLIJN VAN HEEMSTRA

In Search of a Name

IN
LIGHT-YEARS
THERE'S
NO HURRY

Cosmic Perspectives on Everyday Life

MARJOLIJN
VAN HEEMSTRA

Translated by Jonathan Reeder

W. W. NORTON & COMPANY
Celebrating a Century of Independent Publishing

For information about permission to reproduce selections
from this book, write to Permissions, W. W. Norton &
Company, Inc., 500 Fifth Avenue, New York, NY 10110

For information about special discounts for bulk purchases,
please contact W. W. Norton Special Sales at
specialsales@wwnorton.com or 800-233-4830

Manufacturing by Lake Book Manufacturing
Book design by Beth Steidle
Production manager: Julia Druskin

The publisher gratefully acknowledges the support of the
Dutch Foundation for Literature.

Excerpt from "Beyond the Bend in the Road" by Fernando
Pessoa reproduced by permission of Poetry in Translation.
Kline, A. S. (2018), Fernando Pessoa Twenty Poems,
retrieved 6 July 2022 from https://www.poetryintranslation
.com/PITBR/Portuguese/FernandoPessoa.php#anchor_
Toc503461479.

"Evolution" by Rebecca Elson, from the collection
A Responsibility to Awe (2001), reproduced by permission of
Carcanet Press.

ISBN 978-1-324-03569-5

W. W. Norton & Company, Inc.
500 Fifth Avenue, New York, N.Y. 10110
www.wwnorton.com

W. W. Norton & Company Ltd.
15 Carlisle Street, London W1D 3BS

1 2 3 4 5 6 7 8 9 0

For Eyse and Otto

Contents

•

IN

LIGHT-YEARS

THERE'S

NO HURRY

1

●

Longing for an Overview

IT'S THE HOTTEST SUMMER ON RECORD. EVERYTHING feels clammy: the sheets, my pillowcase, David's legs. My phone screen reads 3:32 a.m., which means I've been lying awake for four and a half hours now. A new record. I listen to the fan whirring at top speed, a rhythmic *whoosh whoosh*. It does nothing to cool us down, but at least it gets the air moving and keeps the relentless mosquitoes at bay. David, the man who has been my fiancé for years, shifts restlessly in his sleep. In the next room, our children lie sweating in their beds.

The bedroom window is open, as are all the windows on the open square our house looks out onto. Everyone is hoping to catch some breeze in their stuffy apartment. The last night owls have gone to bed; the early birds haven't woken yet. It is that indeterminate time officially called morning but that still feels like night. An in-between time, the pause between breaths.

I should get up, make use of this scarce moment of quiet. Answer emails, pay bills, write a poem. But instead, I scroll through the stories on my Instagram account. Snippets of anger and indignation alternating with flashes of self-promotion. Poetry and police brutality, rave reviews and ranting op-eds, recipes for vegan kebab followed by a takedown of the patriarchy. I tap hearts and thumbs-ups and share until my head spins from all the battles that need to be fought. We're only halfway through, but this year, 2019, has already been dubbed the Year of Protest: school climate strikes, the yellow vest movement in France, street demonstrations in Hong Kong, Sudan, America. It is also the Year of Divisiveness. A study released this week shows that Europe has never before been so politically polarized.

A front door slams. Bob, probably. My downstairs neighbor leaves at the most ungodly hours for his volunteer work for the Red Cross. Event first aid, visits to the housebound. Bob, the best neighbor I've ever had, but also one I feel awkward around. Because of his overt nostalgia for the good old days in a district now being taken over by people like David and me. Double-income electric-cargo-bike-riding latte drinkers, whose hip coffee joints have elbowed out the Turkish teahouses. Schism at sidewalk level.

3:42 A.M., my smartphone says now, although I could swear it's been at least half an hour since I last looked. *Ping*, a notification. An American activist I follow on Instagram is about to go live. I click and watch: *We live in broken communities*, she says. I toss a hand-clapping emoji into the livestream for that word, *brokenness*. It's exactly what I feel. I swipe through the stories again, open a link to a video about ecocide in the Amazon. Two and a half acres of rainforest are cut down every second, says the voice-over. I count to ten. Twenty-five acres. I try to calculate how many trees, plants, bushes, anthills, bird's nests, fungi, butterflies, and insects are now gone for good. I stop counting, but the seconds tick on. Worlds are being wiped out and I lie here in the dark with a feeling I can only describe as sinking.

SINCE THE BEGINNING OF THIS HEAT WAVE, I have sunk to sleep at night in the bluish light of my screen. I pray for rain, for a puff of wind, for at least a respite from this heat, but the national weather service forecasts no change. Drought, drought, drought.

Our front yard faces south. All day long the sun blasts its searing rays onto the plants and the flagstones. The trees on the square are too far away, their shadows do not reach us. The only thing that shoots up from the sidewalk in front of our house is the streetlamp, whose dirty yellow light shines in past the edge of the curtains at night.

That light has annoyed me for as long as we've lived here. It's like sleeping in a spotlight. I squeeze my eyes shut, search for something in the darkness. A point of view—no, just a view. That's it: I need a view, a vista. I am unable to see past this sweltering night. As though the horizon has snuck up on me, has slowly transformed from a faraway pinstripe into a thick, ugly streak that blocks my view. It seems like the fan has gotten louder. *WHOOSH WHOOSH*. I've lost my view, lost the future, almost as carelessly as one loses the key to a bike. This can't be, it *mustn't* be: to have children, but no clear prospects. What positive story can I tell them? A decent story needs continuity, and yet it's like everything's gradually become disjointed. Trees without a forest, birds without a flock, fish without a school, human- kind boxed in by first impressions. Which, in my case, means: white, left wing, female, a "cargo-bike mom" even without the cargo bike. Where can we go to find something or someone further than ourselves? More than what we seem at first sight?

I SCROLL FURTHER, click on a link to an article about "eco- logical grief" in Greenland, whose people are confronted with the effects of climate change every day and grieve for their lost world. The Australian environmental philosopher Glenn Albrecht has come up with a name for this emo- tion: *solastalgia*, from the Latin *solacium*, meaning solace, and the Greek word for pain, *algia*. Where once you found solace in the landscape, now you find pain.

Maybe that's what's ailing me, but my sinking feel- ing is not limited to the landscape. It is the divisiveness in our neighborhood, our city, the country. Things coming undone. It's the daily stream of apocalyptic news: hous- ing shortages, the burnout epidemic, refugees stuck behind fences, the demise of democracy and of the last northern white rhino. It's my neighbor across the street, who calls the climate crisis a left-wing hoax. This year she plans to take her first airplane trip ever, and suddenly she's sup- posed to be ashamed of it because of the climate kooks like me and our anti-airline-industry petitions. Never mind that I used to breeze from continent to continent on vacation in a time when she and her family spent theirs scrunched into a two-room bungalow.

I sink while my children lie in bed a few feet away, and
I think of a verse from a poem by Fernando Pessoa that a
friend once read to me:

> Beyond the bend in the road
> there may be a well, a castle.
> There may be simply more road.
> I neither know nor ask.
> As long as I'm on the road before the bend
> I simply look at the road before the bend,
> since I can see only the road before the bend.

How wonderful and comforting those words once were
to me, whereas now I get short of breath at the thought of
that road without sight of the horizon. Solastalgia. I want
to climb higher, beyond the bend, look over the valley and
the mountains. I am in search of a road that takes me past
the doomsday scenarios, past the half solutions and the
divisions—to a wide and breathtaking vista.

An image pops into my mind. The "Hubble Ultra Deep
Field," a photo taken by the Hubble telescope from some
340 miles above Earth. The most beautiful vista I know.
It's a photo of a pitch-black background full of illuminated
shards. Light that varies from bright white to a kind of soft
orange, from fluorescent lamp to light bulb. If you look
carefully, the shards all differ in shape. Bizarre spirals,
sparkling snail shells, some whose sharp points stick out at
all points of the compass, like the star of Bethlehem.

The photo is a composite of a series of observations the
telescope made from September 2003 to January 2004.
What look like individual shards are in fact three thousand

galaxies, light-years from us in the constellation Fornax. "We saw to the edge of all there is," Tracy K. Smith wrote in her poem about Hubble's images. "So brutal and alive it seemed to comprehend us back."

If you know what you're looking at, the photo is overwhelming. An unassuming bit of darkness above our heads sparkles and shines. Some of the shards of light are the oldest galaxies we know of, formed in the first five hundred million years after the Big Bang. Galaxies now long extinct, because the light we see had been traveling for billions of years before being picked up by the Hubble telescope. Hubble looks not only forward into space but back in time.

THE FIRST TIME I SAW the Hubble Ultra Deep Field was in 2006 at the Space Expo, the museum adjoining the European Space Agency (ESA) center in Noordwijk, the Netherlands. That year I had undertaken to become the ESA's "house poet," a title of my own design. I'd long been fascinated with outer space. For me, the cosmos was the place for not-knowing, not-being, not-seeing. The awareness of it surrounding us has always made me feel liberated.

I emailed the Space Expo, asking if I might use the museum as a temporary workplace, adding that I had actually wanted to study astronomy in college but that it became religious studies instead, specializing in Islamic mysticism—another route to the unknown. The then director, Rob van den Berg, sent a friendly email back, inviting me to come write in the Space Expo.

That photo—the Hubble Ultra Deep Field—hung near the entrance to the exhibition space, backlit in an otherwise darkened corridor. I spent entire mornings staring at it, mesmerized. The incomprehensibility of thousands of galaxies, hundreds of billions of stars in all.

But no poem came. I got no further than a single stanza. In it, I drew a parallel between the expanse of shards and a Jewish creation myth in which a basin full of light shatters into billions of pieces. According to the myth, all those shards eventually become life. Humans, animals, even words: holy shards of light. Each shard of light, says the myth, is nostalgic for the vat in which everything was once all together. So with every encounter between shards there is a yearning for yet more light, whether those shards are people, animals, plants, or letters.

The broken world is a yearning world.

I SET MY PHONE ASIDE. Out on the square, the first bird announces daybreak with a tentative *prrryit*—a blackbird, a great tit?

Our youngest shouts something in his sleep and I go to the next room to have a look. He's sitting upright in bed, his toddler's face flushed from the heat. "Drink." His brother is in a deep sleep on his sweat-drenched pillow. They look different on these hot nights: bigger, more helpless. When I go downstairs to fetch a bottle from the dishwasher, a plastic tiger on the counter catches my eye—the toy that the younger one has been dragging around with him for weeks now. Of the nine subspecies of tiger, I read recently, three are already extinct and the other six are severely endangered. I pick up the plastic tiger and throw it in the garbage along with the kitchen waste. I don't want a toy in my house that reminds me that my children are growing up with animals that will no longer exist in the wild by the time they're grown. On the way to the stairs I reconsider, go back and dig it out of the garbage bin. An old raisin is stuck to its back. I rinse off the tiger, place it back on the counter. If living tigers are slowly disappearing, the least we can do is hang on to the plastic fake ones.

BACK IN BED, I google the Hubble Ultra Deep Field photo. The light shards are less impressive on a smartphone than on the full-size enlargement at the Space Expo, but they are still beautiful. You wish you could sweep them all together into a glittering beacon in the darkness. According to that Jewish creation myth, this is actually our mission: to reconnect the shards. Words, people, animals, plants—gather everything back together until it makes sense again.

Seen from the perspective of the myth, it's all the same brokenness: climate change, the rift in my neighborhood, the extinction of fungi, trees, tigers. Being separated from things around us that we should be able to connect with: a tree, an ocean, another human being.

Maybe I should finish that poem about the Hubble Ultra Deep Field after all. Lately, it's like I can focus only on the chaos occurring at eye level. All that zooming in, I now realize, has alienated me from something, from the sensation of being part of something bigger, being able to bob around on a rhythm outside myself. I've become too focused on the few square feet around me, on that obsessive "now" of Instagram stories and live blogs and news feeds; I feel bloated from the urgency of it all. I want to shrink, zoom out, find an overview from which I can see things, myself, others, in their proper proportion.

I look back at the shards on my smartphone. The thought of the Hubble telescope is helping to counteract that sinking feeling. Somewhere above Earth floats an enormous eye that sees four billion times better than we do. A collective pupil, built by mutual effort, gazing out into space on our behalf. Can a person be jealous of a telescope? I experience something like envy. To gaze into space, day and night, seems like the ideal remedy for this suffocating anxiety.

I sink slowly into a light sleep with the reassuring image of the Hubble Ultra Deep Field in my mind and a sensation that I am at the start of something. A space journey on Earth. A journey that will take me to the corners of the universe, with both my feet on the ground.

2

•

The Attitude of an Astronaut

IT'S A TROPICAL TUESDAY MORNING IN THE QUIET, DARK hallway of the Space Expo in Noordwijk, and I'm standing face-to-face with the Hubble Ultra Deep Field just as I saw it for the first time, fourteen years ago. Blown up and back-lit so that the galaxies seem to glitter. I open the memo app on my mobile phone, ready to tap in all the grand ideas all this will undoubtedly unleash.

But the longer I look at the forms and colors of all those billions of alien worlds, the emptier my mind becomes. The only thing that comes to mind is a quote by the physicist Brian Cox: "We are the cosmos made conscious." The universe understands itself through us. Staring at the shards, I wonder if it's not also the other way around—if the cosmos is also a means of understanding ourselves.

I walk through the darkened hallway farther into the museum, into a large room divided into separate spaces by low partitions. Synthesized cosmic sounds hum in the background. I am the only one here, and as I amble past the satellite photos of Earth, I feel lighter than I have in days. I glance at my phone; fortunately I still have several hours before David and the children get home.

In the middle of the room is a scale replica of a wing of the International Space Station (ISS). Farther along, behind glass, is the spacesuit worn by the Dutch astronaut André Kuipers, who lived on the ISS from the winter of 2011 through the summer of 2012. I walk past a moonstone, a flag, a small pile of cosmic dust.

Next to a small meteorite, visitors can stick their nose into the opening of a black box and take a whiff of 67P/Churyumov-Gerasimenko, the icy comet on which the space probe *Rosetta* landed in 2014 to collect data for the European Space Agency. Six months after my visit to the Space Expo, analysis of *Rosetta*'s data will clearly show that the comet contains the element phosphorus, a component of DNA and thus a crucial building block for life.

I am moved when I hear this news in the winter of 2020. The relationship between human and comet: easy to overlook, because it extends so far back in the line that led from chaos to consciousness, yet at the same time so fundamental. For it was probably comets like 67P that, when they collided with Earth, left behind our most basic life-building blocks. And without those building blocks there would be no algae, or horses, or bipeds who stick their nose into a black box in a museum in Noordwijk to smell the cosmic rock that carries part of their makeup. The rock, it turns out, smells of ammonia and rotten fish.

As I inhale the rot, I wonder if it wouldn't be a good idea if the story of comet 67P became a fixture of history lessons from now on. If every schoolchild were required to stare for fifteen minutes at *Rosetta*'s grainy images of that alien chunk of rock, to realize that this, too, is part of our origins, that life had already begun before the first organisms arose. A comet as a mutual ancestor of everything alive. Would we look at ourselves differently then?

A school group, eight-year-olds, charges noisily into the room, interrupting my musings. I head to the cafeteria, and a little while later I'm drinking tea with the Space Expo director, Rob van den Berg, while, out on the lawn next to the entrance, a group of children launch homemade hydrogen rockets.

ON THE WALLS OF THE CAFETERIA hang signed portraits of astronauts who have visited this museum. Van den Berg has met many of them personally. I examine the faces, hunting for similarities. Do these spacefarers have a common denominator, a shared characteristic?

Van den Berg considers my question. "Kindness," he says. When you're squashed together into such a small space, the most important thing is to be kind to one another. "And one other thing," he says as his eyes pass along the portraits. Upon returning home after their space missions, almost all of them became Earth activists: ambassadors for sustainability, animal welfare, plastic-free oceans. Having seen Earth from a distance, they agreed, made the greatest impression of all.

I study the faces on the wall. "We are all astronauts on Spaceship Earth," wrote the renowned American architect and futurist Richard Buckminster Fuller in 1968. Back in the 1950s, before humans had breached the atmosphere, he said we were mistreating the world because we didn't see it as a whole. If we could just realize what we are— a tiny round ship in a huge sea of space—we would think differently, live differently. Circular, planetary, no longer as though life were a spring that could be tapped indefinitely. A ship requires maintenance and proper management.

Buckminster Fuller's well-known insight has been constantly replayed over the years, but in this small cafeteria I suddenly get it. From within, as one often does with platitudes.

And all these men on the walls—oh, wait, there's a woman, too—saw in a single glance what Fuller wanted the rest of us to see. Earth as a compact, colorful ship in space.

Van den Berg laughs when I share this thought with him. "You might say that being astronauts in space made them into astronauts on Earth, too." He excuses himself; he has to get back to work.

I check my phone, see that I've just missed the bus to the station, and decide to wait here in the air-conditioned cafeteria for the next one. I get up and read the names under the portraits. According to van den Berg, what they saw up there changed them for good. Could anything from their overwhelming experience be transported back to Earth? I google them, one by one. Two hours later, I'm still sitting at the cafeteria table with my laptop.

I FIND THE ASTRONAUT WHO MET GOD in an empty crater. I find the astronaut who said no one returned from space unchanged. I find the first astronaut who wept on the moon. I find the astronaut who stood in the moondust, looked out whence he came, and said, "We went to discover the moon, and we discovered the Earth."

I find a video of the Dutch astronaut Wubbo Ockels, who on his deathbed recorded a video message to warn his fellow earthlings that we are not "aware of the danger in which we live. . . . Our earth has cancer. I have cancer, too. And most people with cancer, they die." His English sounds wooden and Dutch, which somehow makes his video message seem even more vulnerable. Pale, emotional, and out of breath, he beseeches humankind to see the planet through the eyes of an astronaut. From a distance. To see it as an entity to which we are inextricably bound.

It is a heartbreaking scene: the dying astronaut who, in his last breath, tries to save a planet from self-destruction. But near the end of his message, hope shines through. We can do it, he says, his voice breaking. If you have "the insight and the attitude of an astronaut, you start to love the earth in a way other people can't, and if you really love something, you don't want to lose it."

When the video is finished, I click Replay. *The attitude of an astronaut*. For a moment I feel uplifted, I feel myself, sitting in this cafeteria, floating through a solar system, a flash of the enormous space surrounding me. Overview.

I scroll further, click, and listen to the space travelers who, once outside the atmosphere, were overcome by love for the planet, by the urge to protect it, by a tremendous sensation of wholeness.

"It does something to you," said Edgar Mitchell in 1971 upon returning from the Apollo 14 moon mission. He experienced a powerful awakening up there. An overwhelming awareness of the interconnectedness of life and an intense dissatisfaction with the way we treat that life. "You want to grab a politician by the scruff of the neck and drag him a quarter of a million miles out and say, 'Look at that, you son of a bitch.' "

On his way back from the moon, Mitchell felt a connection "on a subatomic level" with his crewmates, with the capsule in which they traveled, and with the darkness outside it. Later, he would say that he appreciated then how everything was born in the heat of an ancient star. How everything is essentially extraterrestrial, and through cosmic collisions ended up on Earth.

Differences cease to exist, the space tourist Anousheh Ansari told a journalist after returning to Earth. Ansari, a wealthy entrepreneur, flew to the International Space Station in 2006 as a civilian passenger, together with two astronauts, on a Russian rocket. The view from space, she said, was life-changing. "When you see it from that angle, you cannot think of your home or your country. All you can see is one Earth."

APPARENTLY THERE IS AN OFFICIAL TERM for this transformative experience in space: the "overview effect," coined by the American author Frank White. In the short documentary *Overview*, White explains that he laid statements by thirty-one astronauts side by side, and discovered that the core of these observations was a cognitive shift when viewing the earth from space.

White's research dates from the 1980s, two decades after the famous "space race" between the Americans and the Soviets and the first moon missions. In the early years of piloted space exploration, there appeared to be little attention to this "soft" side of space travel. At least, I can't find any articles or research about it. Perhaps it doesn't tally with the image of the cool, collected astronaut who plants flags and footprints in the moondust.

White's research sums up the overlapping elements of this cognitive shift, and indeed, they bear little resemblance to rockets and the competition for outer space: love for planet Earth, a desire to protect it, a sensation of connectedness to everything that is alive. White concluded that this cognitive shift, in many cases, was long-lasting.

This is exactly the distance, the overview, I've been looking for! To reset my perception of my relationship to everything I'm part of. And, especially, a lasting experience that gives me some room, counteracting the claustrophobic anxiety that's been haunting me for weeks now. The attitude of an astronaut.

Eager to learn more, I order a coffee from the cafeteria lady and return to my laptop. Online I find a follow-up study of the overview effect, this one conducted by researchers at the University of Pennsylvania. Analyzing preexisting interviews with astronauts, a team of research psychologists studied the exact circumstances under which the effect occurs. Strangely enough, it is the enormous physical distance from Earth that seems to trigger the feelings of emotional closeness to it. Apparently, at a certain point you are so far away from something that you see it in a completely different light.

Additionally, the researchers established that gazing at the Earth can have therapeutic effects. It provides something in chronically short supply these days: *awe*. The word *awe* itself is a sort of visual onomatopoeia, making your jaw drop automatically when spoken out loud.

Further along in the study, I encounter the word *transcendent*. There are many parallels between the astronauts' descriptions and those given by people who have undergone a mystical experience. Astronauts: a bunch of mystics circling the planet, completely detached and yet nearby.

The woman from the cafeteria comes to tell me they're about to close. I nod; I should go, I should have gone a long time ago, I still have to do the shopping before David and the boys get home.

I reread my notes on the Penn study one last time. Finding proximity by creating distance. Zooming out to understand what's happening in front of your nose. Detaching yourself to experience how profoundly attached you are to the place you inhabit.

The astronauts interviewed in the documentary *Overview* see it as their mission to share their insights with the rest of the earth's inhabitants. But their spectacular flight is a far cry from my daily spin around the sun. Could I be as open to that experience as a trained astronaut during their long-awaited journey into space?

Soon I'll be back in the sizzling heat, watching our small front yard wilt, day by arid day. Soon the world will be broken back into pieces: Instagram profiles, news sites, our neighborhood square. Soon I'll be sitting there again among the plastic dinosaurs, growling, "*Whaaaahaaa*, I'm a deinonychus!" And when the eldest shouts that he wants to be a Parasaurolophus when he grows up, and the youngest a T-Rex, I'll again have to bite my tongue so as not to tell them that we are in fact the dinosaurs of our time: a species on the verge of extinction.

I PACK UP MY THINGS and walk to the bus stop in the harsh afternoon light. Next to the Space Expo is the European Space Research and Technology Centre, ESTEC, the technological heart of the ESA, where space missions are conceived and managed, and where the satellites and instruments for those missions are tested. The grounds are surrounded by a tall, metal fence.

I've been inside it a few times, at open days and news conferences I attended—out of curiosity, without reporting a single word. I didn't dare, afraid of advertising my complete ignorance of mechanics and physics; I can still kick myself for having pursued the humanities path in high school. The fence suddenly strikes me as strange: barbed wire separating the museum and research facilities. As though culture and science mustn't contaminate each other.

There's the bus. The seats at the rear are already taken by ESTEC personnel, identifiable by their lanyards and ID cards. As we pull out, the sky clouds over and a light rain starts to fall. At last. All the way along that boring stretch between Noordwijk and Leiden, I keep thinking about that paradox from the University of Pennsylvania study: distance engenders proximity.

In opening his famous 2005 college commencement speech "This Is Water," the American essayist David Foster Wallace introduces a pair of fish swimming along. They happen to meet a third fish swimming in the opposite direction, who nods and greets them: "Morning, boys. How's the water?" After swimming on for a bit, one of them looks over at the other and says, "What's water?"

Wallace's point is that the most obvious truths are often the ones that are hardest to see and talk about. Invisible behind the camouflage of the everyday. Familiarity breeds ignorance.

Up in space, the astronauts saw anew, with a fresh gaze, something that had been surrounding them all along. The remoteness sharpened their view; the distance allowed them to suddenly comprehend where they had been living all that time. We are all astronauts, but we keep forgetting it. We inhabit a small planet at the edge of a galaxy we call the Milky Way; every day, we hurtle in our orbit through the universe.

The bus weaves its way to Leiden station, and I wonder to myself how I can achieve the overview effect on a complicated neighborhood square on earth. The word *overview* bounces around in my head until late in the evening. Overview. The space of that capital *O*—like the eye of a telescope.

3

•

Earthgazing as Therapy

I LEAN OVER A LOW PLEXIGLAS WALL AND WATCH THE earth rotate under my feet. I try to make out the continents, the pale patch of the Arabian Peninsula, green-splotched South America. But what I see most is sea blue, alternating with the milky white of passing clouds. Here in the Netherlands we're rotating faster than I would have guessed: about 650 miles per hour, every hour of the day, every day of the year. And that's only the spin of the earth around its own axis. Add to that our 67,000-mile-per-hour revolution around the sun, which in turn whips around the center of the Milky Way at a speed of about 490,000 miles per hour. I press my knees against the Plexiglas panel, try to suppress the dizziness as the Pacific Ocean sweeps underneath me.

I'm looking at the view from the International Space Station, some 250 miles above the planet, projected onto a giant screen below. I am in Kerkrade, at the southernmost tip of the Netherlands, in the Columbus Earth Center, a museum devoted entirely to the astronaut's view of Earth.

This seems like a logical place to start my search for an equivalent of the overview effect. To see what astronauts see, feel what they feel. So this morning I took the train to the south of the country, and here I am staring down into this cylinder. Once the dizziness has subsided, the steady pace of Earth and the clouds has a soothing effect. It's like looking at the ocean. We all have a "blue mind," the marine biologist Wallace J. Nichols once said. The sight of water, essential for life, puts people at ease.

I give the watery sphere one last look. It's beautiful, but I'm not overwhelmed the way the astronauts were. I try to imagine what it would feel like to see this for the very first time. But I'm unable to separate it from the countless Earth-image mouse pads, coffee mugs, posters, and T-shirts I've known my whole life. The most reproduced image on Earth is perhaps a photograph of Earth itself.

THE FRENCH PHILOSOPHER BRUNO LATOUR holds that the image of our planet seen from space offers a misleading idea of humanity's position. The photo wasn't taken from space, he says. It was taken from inside the cramped, noisy capsule of a rocket, a place where no human can survive without artificial life support.

For Latour, that famous Blue Marble photo of Earth taken from space is too sentimental and therefore skews our worldview. We mustn't look at the world from without, but from within. We are here, not there.

As I walk away from the Plexiglas, behind me I hear a new group of visitors shuffle in, ready to gaze down at themselves. Bruno Latour's comment nags at me. For him, looking at Earth from outer space constitutes unhealthy escapism. Zooming out to avoid having to zoom in. Is that what I'm doing? Dodging reality?

Reading most astronauts' statements, you can only conclude, contrary to what Latour says, that there is no escape. Or that if there is, it boomerangs right back at you. Those astronauts, once back home, commit themselves more than ever to the planet they temporarily left.

Well, most of them do. Here and there, you do find the story of spacefarers who were apparently immune to the overview effect. In an episode of the podcast *This American Life*, the astronaut Frank Borman recounts how indifferent he was to the sight of earthrise from his *Apollo 8* capsule. Back home, he didn't even talk about it with his wife and children. "It was more important to see the boys and see her. . . . We [just] got right back to the nitty-gritty's" of everyday life.

Compare that to the effect earthrise had on Borman's two fellow astronauts, Jim Lovell and Bill Anders. The view of Earth reminded Lovell of all the times he had heard people say they hoped to go to heaven when they died. But heaven, he realized up in space, is where we were born. It is that tiny planet that provides us with everything we need to thrive. And for Anders, "borders that once rendered division vanished. All of humanity appeared joined together on this glorious-but-fragile sphere." Maybe, in addition to seeing it, it's also a matter of *wanting* to see it.

WHEN I GET BACK FROM KERKRADE, my neighbor Bob is furtively watering the plants in defiance of a nationwide sprinkler ban. "Otherwise they won't survive," he says, nodding at the lavender, his favorite.

I watch as Bob pinches dead leaves from the dry plants. His perpetually tanned limbs, his combed-back hair. He has set a dish of water next to the lavender for the birds. Bob tells me he used to have a small rose garden here. One day, a landscaping firm came and cut it down. Too many front yards in the neighborhood were being neglected, so the city council decided to outsource the grounds maintenance. Gone were the roses, and in their place came the rows of rugged, vandal-proof hedges that were there when David and I bought the upstairs apartment.

Shortly after we moved in, Bob offered to share his front yard with us. There was nothing about it in our contract, but we enthusiastically accepted his offer and since then we have replaced the hedges, bit by bit, with plants of our choice. At first we had to protect them from the hedge clippers, but now the landscapers have caught on and skip us on their monthly rounds.

Bob asks where I've been today, and I mumble something about working on a report for work. Here among the withering garden plants, I can't find the right words to describe what I'm looking for. Connection, overview, the attitude of an astronaut: it all sounds so remote from this patch of yard, from this square, from Bob in his cutoff jeans and white T-shirt.

As I walk inside, Latour's remark is stuck on a loop in my mind. *We're here, not there.* But aren't we in both places? Here *and* there? Why choose? From within or from without?

The house is quiet. Evidence of this morning's rush is

everywhere: an upset drinking mug, stray socks, the crumb-strewn breakfast table. I flip open my laptop and turn on the TEDx Talk I had googled on my way home. I don't much like TED Talks—all that pacing back and forth, the obligatory punch lines—but this is one speech I simply cannot skip: "The Therapeutic Value of the Overview Effect."

ON A SMALL STAGE SOMEWHERE IN LONDON, the psycho-therapist Annahita Nezami (dark medium-length hair, a nervous laugh) tells an audience how her quest for a comprehensive therapy led her beyond the atmosphere. She saw that the way we live together feeds our insecurity and greed, she says. Fear and negativity circulate throughout modern society in thousands of forms, causing depression and loneliness. As a psychologist, Nezami explains, you can treat all those symptoms individually, but you can also ask why we've run up against this wall en masse. She pauses, looks into the auditorium. "Brokenness," I mumble in our silent living room.

We face deep-seated alienation on all fronts, Nezami says. Lost connections with one another, with our jobs, with the place where we live. Psychologists shouldn't only treat individual patients, she says, but should heal society as a whole, too. "A bit of a tall order," she laughs.

Where do you start looking for a solution to global alienation? Maybe in its opposite: global connection. For her dissertation, Nezami scoured scientific theories about how people experience cohesiveness. While doing so, she stumbled upon Frank White's study and the testimony of those thirty astronauts who, far outside the atmosphere, felt deeply connected to the entire web of life.

Ha, Frank White! This week I revisited his documentary about the overview effect. I hoped to catch something
new, something I might have missed the first time, a hand-
book for those of us within the atmosphere wanting to feel
what astronauts felt outside it.

I hadn't missed anything. The film is a paean to the attitude of an astronaut, but offers no suggestions for earthlings.

Nezami does. To better understand the effect, she says,
she conducted in-depth interviews with seven astronauts.
These interviews more or less corroborated the conclusions White and the subsequent University of Pennsylvania
researchers reached. The astronauts she spoke to experienced cohesion where they first saw division, and were
overcome by a sensation of belonging—to humanity, the
forests, the wind, the lightning, everything that makes
Earth Earth. And most of them still feel it to this day.

These astronauts were also part of White's study. So,
not a huge test group. Scientifically speaking, the study of
the overview effect is a winding path, not yet paved with
heaps of hard facts—yet it still feels like the route I've been
looking for.

Nezami, by the way, has her doubts about the effect: one
look at Earth from space, she says, is not really enough to
achieve a full overview effect. Her research reveals that the
more the astronauts looked at their home planet, the more
intense the experience. I think of this morning in Kerkrade,
where I gave up after hardly fifteen minutes. Not exactly
the attitude of an astronaut.

And then, more nuance. As the end of Nezami's talk
nears, a video plays on the screen behind her, panning
slowly over the surface of a darkened earth. The aurora

shimmers along the horizon, then gives way to the gleam of city lights. I think of how we associate the experience of astronauts with the famous Blue Marble photograph of Earth as a fragile little ball hanging in the darkness, the only traces of humankind being damage and destruction. But at night, the impression our planet offers from space is completely different. I see how human presence, all those countless twinkling lights, gives the nighttime view of earth a lively, vital appearance. People are not only destroyers of the ecosystem, but also illuminators of the night.

And then Nezami says just what I've been waiting to hear. Most of us will never leave the atmosphere, but "I really want to try to bring this experience down to all of us," she says. I turn up the volume on my laptop so as not to miss a word as Nezami walks past the iconic TEDx letters and explains a virtual reality program she's working on. A VR experience that allows you, in multiple sessions, to hover in space like an astronaut, with a view of Earth, both at daytime and at night. Earthgazing as therapy.

I LOOK UP NEZAMI'S EMAIL ADDRESS while the London audience applauds and cheers. I want to don those VR goggles, gaze therapeutically at Earth. But when I speak to her a few days later, she tells me the project is still in its infancy. They need more research, a pilot, a test group; it will be months before the first phase of the project is ready to be implemented. Developing the goggle software turned out to be complicated. "We're still experimenting with how we can implement language and music to bring the viewer into just the right astronaut-like state of being."

Outside, Bob shouts to someone. I walk to the window,

phone to my ear, and while Nezami continues speaking about how to turn the earthbound into astronauts, I see John, a neighbor from a few doors down, step into the front yard. His round belly hovers like a planet above his legs. Yellowish with a pink glow in the sunlight. On hot days like this he goes shirtless, like so many men in this neighborhood, men who spend their summer on chairs out front, their big bellies a solar system of planets around the square.

Bob looks up, sees me standing there, waves. I see myself through his eyes. A feverishly telephoning mother, her children at day care, still an interloper after seven years in the neighborhood, while he belongs here. Bob, who suffers from the gentrification I embody, whom I see every morning in the front yard but never in the café where I drink my latte or at the hip art-house movie theater that just opened up nearby. How would I explain this conversation to him? A telephone consultation with a London psychotherapist who tells me how seeing Earth like an astronaut is an antidote for doomsday thoughts. Just as Nezami pauses, Bob walks back inside, and in a flash I imagine he picks up his phone on the kitchen table to take over the call to London. What would he have to say about this self-help project, about my attempt to reset my understanding of the world via the universe? What would his remedy against divisiveness be? He comes back outside carrying a shovel, and sets about moving a thirsty plant to a shady spot.

I ASK ANNAHITA NEZAMI whether she ever doubts her mission to use the attitude of an astronaut to cure society. She laughs. "Yes and no. I think today's problems are so comprehensive that they'll require a comprehensive therapy."

I stifle a sigh. Against my better judgment, I had hoped for a ready-made key to the overview effect.

"It's a mad undertaking," Nezami continues. "But the way we live today is at least as crazy. We treat psychological ailments with pills and therapy targeted at the individual. But what if the problems are bigger than the individual? If those ailments are the logical consequence of our relationship to the world?"

John walks back home, Bob coils up the garden hose. A flock of ring-necked parakeets swarms above the square, drawing a bright green stripe across the houses. Nezami's last words comfort me.

This quest of mine feels idiotic. But with David Foster Wallace's fish in mind, maybe "idiotic" is just what I need. A path that leads me away from the route of reason, from what I know and understand, away from the routine, from the water we are swimming in. I need something to pry me loose from my daily patterns, from my anxiety, so that I can see where I am. Earth. Space.

I ask Nezami what advice she would offer someone who lives in a second-story apartment in the city and is looking to feel what astronauts feel. After a moment's silence, she says, "A condition for the overview effect is awe. Looking at Earth from space is comparable to experiencing a breathtaking landscape in the mountains or the forest. But in a city? I don't know. In a city, there's not much that's bigger than ourselves. The only thing I can think of is to look up. Light in the darkness, the starry sky."

4

•

Spacefarers without Stars

IT'S LATE IN THE EVENING; THE SKY IS CLEAR AND THE heat has subsided. David is sitting with friends on the terrace of our new neighborhood café at the far end of the square. The café that replaced a bar where drugs were found during a police raid. Half the square is delighted to have kimchi grilled-cheese sandwiches and sweet-potato fries within reach, the other half can't afford "that pricey stuff" and gives it a wide berth.

"The neighborhood's going to the dogs," one neighbor grumbled recently as she eyed the new outdoor café. "First, Kaddour left, and now this." I nodded awkwardly. Of course she's already seen me sitting at that café, but I miss Kaddour, too, the butcher who recently moved to larger premises up the road. It was one of the few places in the area where everyone, veterans and newcomers, met in the space of a few square yards. And it was the only butcher shop that I, a vegetarian, frequented. For the olive oil and the nuts, but also because the pleasant atmosphere seemed to neutralize all the differences among us.

The late summer sun has set, but the warmth still hangs over the square. It took the children forever to fall asleep in their stuffy room, but now I can sit with a can of cola on the bench in the front yard with my face turned toward the night sky. Against my better judgment, I had hoped for a little awe, but instead I feel dismay. I count all of thirteen stars and one planet that turns out to be an airplane.

This week I started reading Paul Bogard's book *The End of Night*, about the importance of nighttime darkness. It features a quote by the American astronomer Bob Berman, who reckons you need at least 450 stars for your jaw to drop open when you look up. "There's a certain tipping point," Berman says, "where people will look and there will be that planetarium view. And now you're touching that ancient core, whether it's collective memories or genetic memories or something else from way back before we were even human. So you gotta get *that*, and anything short of that doesn't do it."

AS I DOWN THE LAST SWIG OF COLA before going back inside, Bob appears at his front door. He nods at the hard wooden bench. "Need a cushion?" I shake my head. He stays in the doorway, hands on his hips, his ready-for-a-chat posture. I ask if he used to see more stars than this. He nods.

"You used to be able to see all the constellations."

We gaze upward, blinking at the light of the streetlamp.

"Not much left," he says dryly.

John walks up and joins the conversation. Says he's

interested in the stars, in astrology. He asks me my sign and nods approvingly when I tell him Aquarius.

"I thought so." His daughter is a Taurus, he says, which explains why he never sees her anymore.

I point to the streetlamp and ask if he doesn't think it's a pity that it blocks out the stars. John shrugs his shoulders. "It used to be darker, but then it was a hangout for riffraff. They hassled the neighborhood kids, it wasn't safe at night. So we asked the city for more lighting."

"And?" I ask. "Did it get safer?"

John shakes his head. "But it wasn't because of the dark. It was because of the people." When I suggest that we could do with a little less light now, he looks at me askance. "You don't know where you're living."

Back inside, his comment has stuck with me. It's true, I don't know where I live. But isn't it precisely because of all that light? Who does know where we live, in relation to Venus or the Big Dipper? Who navigates by the stars anymore? Our natural signposting has more or less vanished.

A FEW DAYS LATER I drive with my eldest son to an observatory in the dunes. A place where they give talks on the universe but also, like tonight, let you look through a huge telescope. This is past his usual bedtime, so I've put him in his pajamas in case he falls asleep in the car. But he's wide awake on the back seat, determined not to miss a moment of this "night expedition."

Even through the telescope, however, there's not much to see except the glimmer of a crescent moon. There is an attendant, a woman with a gray bun who teaches my son a depressing little rhyme: *For every star that you see shine, there are unseen another nine.* She answers my grimace with a shrug. "We live in one of the most overlit countries in the world."

In the old days, she tells me, there were astronomy bulletins, akin to the weather report. You could read in the newspaper what would be visible in the sky that night. People used to be able to tell east from west from the constellations. She could see the Milky Way from her balcony on the outskirts of Amsterdam. She would spend whole nights with her father, looking up at the stars above the sleepy postwar city. "Generational amnesia," she says. "Every generation grows up with fewer stars than the previous one, but has no memory of how it used to be. So nobody misses them. Ignorance is bliss."

On the drive home, the sliver of moon hangs above the fields. In the back seat, my son has given in to sleep. I look in the rearview mirror: in the light of the streetlamps he looks pale, translucent, almost otherworldly. I am reminded of that Jewish myth of the vat of light that broke into shards and yearns for the wholeness it once had. The fields give way to industrial zones, and we drive through a blindingly overlit landscape into the city, two loose shards on their way home.

DAVID AND THE YOUNGEST are already asleep when we get home. After putting the eldest to bed, I scroll through "The New World Atlas of Artificial Night Sky Brightness," a large-scale study of the increase in light pollution. The online map of Europe shows Holland's "Randstad"—the metropolitan belt stretching from Amsterdam to Rotterdam—glowing like the tip of a cigarette. An alarming swath of dark red speckled with, if you look closely, light pink, the color of peeling skin—the color code for the highest level of light pollution on earth.

In the Netherlands, says the atlas, this is due to a combination of factors: the well-nigh endless stretch of greenhouses in the west of the country, the harbors, the reflection of all that water we have here, the population density. Zoom out to the world map, and the Netherlands is a shining beacon of artificial light, which explains the dearth of stars. We are one of the top five most light-polluted places on the planet. This excess of light makes our eyes less able to accommodate darkness. Moreover, the light pollution creates a dome of reflecting dust particles in the atmosphere, which in turn impedes our view of the heavens.

According to the atlas, 99 percent of Europeans and North Americans live under light-polluted skies. Eighty percent of the children born on these continents today will never see the Milky Way in its full glory. The remaining continents, too, are rapidly losing their natural night light. If Earth is a spaceship, then it is a ship without a view and I am a spacefarer without a starry sky.

Before going to bed, I open the blinds and look outside one last time. The harsh light of the streetlamp, as usual, blocks my view. I miss the night I never had. The woman at the observatory soberly summed up generational amnesia by saying "Ignorance is bliss."

Well, I'm not ignorant, and it's not bliss. I long for a view, grandeur, awe.

5

•

Light and Night

NIGHT IS FALLING JUST AS ECOLOGIST KAMIEL SPOELSTRA
and I enter the woods in the Utrechtse Heuvelrug, a national
park nestled between the cities of Utrecht and Amersfoort.
From the pocket of his windbreaker comes the crackling
of a bat detector, which will go off a lot during our walk.
With every click of the machine comes an explanation of
what the radar has picked up. Low flyers, high flyers, timid
bats, brazen ones.

It is a soundscape that goes in one ear and out the other,
because I don't know what to listen for. The only thing I
recognize for sure is the freeway off in the distance.

Spoelstra is clearly in his element on these sandy, shadowy forest paths; I tag along somewhat awkwardly.

Officially it's still summer, but it feels like autumn,
there's a chill in the air, I should have dressed more warmly.
This late in the day, we're the only hikers here, and I feel
a vague anxiety well up at the thought of the approaching

darkness. Spoelstra isn't bothered in the least. He loves the night, he says; as a child he used to sneak outside to watch animals while everyone else slept.

We are headed to a sand quarry at the edge of the woods that looks out over a valley of light pollution. As the colors of daytime gradually recede, birdsongs fill the air. My guide comments on these, too.

"Do you hear that? That chirring!" (A robin.)

"Hah, he's imitating a buzzard!" (A thrush.)

"He's just coming to check us out." (A tawny owl.)

THAT DISHEARTENING DITTY the observatory attendant told us has been on my mind for weeks now. *For every star that you see shine, there are unseen another nine.* If I can't experience the grandeur of the night sky directly, then why not indirectly? You can also get to know something by studying its absence, by tracing the contours of the deficit.

That's how I came to meet Kamiel Spoelstra. Eight years ago, Spoelstra instigated a large-scale study of light pollution in the Netherlands on behalf of the Netherlands Institute of Ecology. He sums up his findings as we walk down the path.

Birds like the great tit tend to breed prematurely if they live close to artificial light, which means there's not enough food available when their chicks are born. Even at very low light levels, far away from a streetlamp, birds still sleep restlessly. Some bats, afraid of light, lose territory, as do wood mice and various types of martens.

Other bats, conversely, benefit from the streetlamps because the light attracts insects. "The insects fly around in circles until they either drop dead or get eaten. The severe drop in insect populations is to a large extent caused by artificial light."

If artificial light increases globally at the present rate, Spoelstra fears it could contribute to species loss. "Sea turtles, for instance. They lay their eggs on overlit beaches, the hatchlings become disoriented, and are thus easy prey for crabs."

Light pollution is finally getting some attention, Spoelstra says, but still not enough. One of the reasons is that no one does what we're doing now: go out at night, seek out darkness. "Darkness has been banished from our lives. There's even an absurd Dutch law that forbids going into forests after dark. We'll no longer know what we're missing when night disappears." At times Spoelstra feels like he's a man with a mission: he wishes he could drag people into the woods at night. "But at heart, I'm a scientist, not an activist. And the problem with the loss of that nocturnal experience is that it can't be quantified."

I tell him about my visit to the Erasmus Medical Center in Rotterdam a few weeks earlier, where I went to ask about the effect of night deprivation on our biorhythms. They took me to a small lab where a colony of mice were doing shift work. They're kept awake when they would normally sleep, and vice versa. A metal rod rotates slowly through their cage, rousing the animals out of their sleep at unusual times.

Chronobiologists studied the physical effect of the mice's disturbed circadian rhythms. There, in that bright, sterile room suffused with the smell of sawdust and mouse urine, I realized I had found myself in a scale model of our out-of-control twenty-four-hour economy.

A nonstop nowness that exhausts our system. The absence of absence.

Spoelstra nods; he's familiar with that study. Laboratory tests on mice have shown that circadian disruptions correlate with a greater incidence of diabetes, cardiovascular disease, and perhaps even cancer. Tumors develop earlier and grow more quickly when the body's biological clock is tampered with.

You don't even have to work shifts to suffer from chronic circadian disruption. Even those who work regular office hours feel the effects of the twenty-four-hour economy. We all go to bed too late. Chronobiologists say we would be much better off if we lived according to our natural rhythm. With less light in the darkness.

IT SOUNDS SO ROMANTIC. Embracing the night. But I notice the slight panic that spreads through my body as dusk falls. The silhouettes of the trees and bushes, the spooky way the path is illuminated, the distant grunt of a deer.

It is beautiful, but in a disturbing way. A falling branch makes the hairs on my neck stand on end. I might be worried about light pollution, but in a world without artificial light I probably wouldn't dare leave the house at night.

"There's no evidence that light really makes us safer," Spoelstra says when I share my thoughts with him. "Research has even disproved the link between light and safety."

He refers to a study that showed that well-lit neighborhoods experienced more criminal incidents than poorly lit ones, simply because there was more activity, and thus more undesirable activity. I am reminded of the streetlamps on my square.

I look it up when I get home. The study, *Outdoor Lighting and Crime*, drew on data from Canada, America, and England. The researchers concluded that better lighting provides more safety only in combination with social control. Lighting on its own benefits only criminals: it's easier to break in when you can see what you're doing.

Likewise, in traffic, more light does not always mean safer roads. If the traffic situation is manageable without overhead lighting, it's better to leave it that way. People are more alert when they have to pay attention, which in turn means fewer accidents.

But even with all this information, I'm not able to reason away my deep-seated fear of the dark. It has nothing to do with logic and everything to do with demons, witches, and horror films. "Fear of the dark is essentially unspecific; like darkness itself, it is formless, engulfing, full of menace, full of death," wrote the poet and author Al Alvarez in his book *Night*. That fear keeps me from looking into the distance, makes me long for what I know.

"SOMETIMES, THE MORE LIGHT YOU HAVE, the less you see," Spoelstra ventures, in a renewed attempt to convince me. He shines his flashlight on a few nearby trees. "You see these trees really clearly now, but the light obliterates everything around them. Light enlarges your surroundings on the small scale and reduces them on the large scale."

IN DOUGLAS ADAMS'S SCIENCE-FICTION NOVEL *The Hitchhiker's Guide to the Galaxy*, the planet Krikkit is enveloped in a permanent dust cloud, so the inhabitants never see the stars. One day the cloud dissipates and the Krikkiters discover they are not alone in the universe. Shocked by the realization that the universe does not revolve around them, they can think of only one response: declare war on the cosmos.

What does the lack of a starry night do to the inhabitants of Earth? Does our narrowed horizon reduce us to small-scale thinkers? To people who can no longer relate to the immensity of our surroundings, because our window to the universe is constantly fogged over?

Spoelstra nods in agreement.

We turn down a path to the sand quarry, our destination. I'm surprised at how quickly my eyes have adjusted to the dark. And slowly, I start to feel what Spoelstra means about nature and nighttime. A world of shadows that stimulates all your senses, slows your pace, opens your ears.

In some cultures, darkness is a time for healing. Exactly what we fear—lying awake at night brooding—is encouraged, for only at night can you see the full scale of your pain, doubt, grief. And only then, when you see it, can you deal with it. And whoever doesn't deal with things will not be transformed. In *The End of Night*, Paul Bogard quotes a member of the Iroquois tribe who describes the spiritual importance of night:

"WE HAVE THE NIGHT so the earth can rest. . . . We have the night so we can release our spirits to travel across space and time. . . . It is by night that we cross into other worlds, other times past and future. Only through dreaming are we able to make peace with the fact that our existence is more than we see during the day. We are never alone; nor are we restricted by the body as long as we use the night to see our place in the right perspective."

COULD IT BE that our lost view of the stars has blocked out our view of the future? That our everyday hustle and bustle, the stress, the artificial light, are so ever present that we, slowly but surely, can't imagine anything except this frenetic, infinitely recurring here and now? A constant battle for those few square feet in our field of vision? Could it be that we experience more brokenness because there's no more darkness into which our differences can dissolve? Darkness in which borders become visible, where we can breathe, and where we can step back from everything we so resolutely stand for in daylight?

The path veers off and then we are standing at the edge of the vale. Below us lie the towns of Veenendaal and Ede—two beacons of light. The A12 freeway, a river of automobile headlights, glistens between them. Above the next hill over, the clouds glow pink—the reflection of the city of Arnhem. After an hour in the dark, the sight of so much light is rather shocking.

I ask Spoelstra how detrimental this much light is for the animals in the woods. "Very," he says. "The great tits in our research became restless with even a small amount of light at night. So you can imagine the effect of this kind of illumination."

I LOOK AT THE SKY ABOVE THE VALLEY. It's mostly clear, but still I can't make out many stars. Certainly not the 450 that the astronomer Bob Berman says you need in order to be overwhelmed by the beauty of the night sky.

Spoelstra and I look in silence at the glow of the two nearby towns. Even the bat detector has gone silent. The ecologist sighs. I ask him what can be done.

"We need legislation," he says, "like in Slovenia, where there's a maximum percentage of light allowed to shine up into the sky. We need to make better choices about the color of light we use. Roughly speaking, the bluer the light, the more damage it does. Red light has much less effect on the environment. Turning off the lights at home helps, but the real benefits lie in cutting back on lighted public spaces and the greenhouses. Local councils need to be made aware that a lot of public lighting is superfluous."

I think of the work of the French photographer Thierry Cohen, who digitally combines photos of urban skylines with a clear night sky, without light pollution. They are breathtaking cityscapes in which the buildings melt into the stars.

"Perhaps the biggest problem," Spoelstra says, "is that light is so cheap. We thought LED lighting would be the answer because, since the light source is so focused, the lamp doesn't scatter light as much. But all we've done is put up more of them, because they're so inexpensive. There's no financial incentive to turn off the lights."

Spoelstra has three young children, who have never seen the Milky Way. "In the summer, we go camping in dark places, but then the nights are so short that they're asleep before the stars come out. And in the winter we're here, where you can't see the Milky Way."

The bat detector picks up the regular chirp of a dwarf bat, one species of bat that thrives on light, feeding on the insects the light attracts. We stand at the edge of the sand quarry for a moment before heading back into the woods. "A crying shame," Spoelstra mumbles.

WHEN I GET HOME, the youngest is on the sofa watching *Dinotrux*. David shrugs at my questioning look. "He wouldn't sleep." Glued to the TV screen, my son does not seem to notice I've come in. At the beginning of the boys' dinosaur obsession, I looked up what it is about extinct reptiles that fascinates children so much. I found dozens of explanations, the best one being that for small children, dinosaurs represent grown-ups: really old and really big. But when I see my sons staring at the screen, I sometimes think that the animals trigger something in their little reptilian brain, a prehistoric kinship. I lure him with a plastic dinosaur egg from the sofa to his bed, where he finally falls asleep. A small, warm predator with a baby T-Rex clutched in his fist.

I think back on the evening's walk. How is it possible that the disapperance of the night could go unnoticed? While David turns off the downstairs lights, I google "World without stars" in bed. The first hit is a 2013 thriller, *Sterrenmoord* (Murdered stars), by the Dutch philosopher Govert Derix. In a YouTube video, Derix explains his fascination with the night and the tragedy of the vanishing darkness.

What does a philosopher have to say about our loss of the night, I ask him a few days later over the phone. Derix says that we are losing part of our history. "Mankind has been looking upwards since the beginning of his existence. The constellations represent some of our oldest legends. Thousands of years ago, people recognized patterns in the night sky and drew imaginary lines between the points of light. The universe is one of the primal sources of the human imagination. And of science. It's no coincidence that the word *cosmos* means 'order.' " Now, he says, we're becoming alienated from this source. "For thousands of years, our species has been looking up into the sky, found solace and wonder in the stars. And now, the thought that we're the first generations to cut ourselves off from that view . . ." Connection is what it's all about, says Derix. With nature, with history, with all those people before our time who looked up at those same stars. That awareness makes us important and negligible at the same time. "Pessoa wrote some beautiful poetry about it," he says just before he hangs up. "About the greatness and the nothingness of the universe."

Afterward, I find the poem. Describing the wide view of the sky he sees from his home atop a hill in a small village, Pessoa observes that our size is defined not by our physical height but by the scope of our view. "I am the size of what I see," he writes. Thus life in cities, where buildings obscure the open sky, diminishes us by limiting the range of our vision. And, the poet concludes, we are impoverished, "for our only wealth is seeing."

6

•

Cosmological Awareness

FOR TWO MONTHS NOW I'VE BEEN MEANDERING AMONG
initiatives, studies, people, places, books, films, photos,
and thoughts that give me the feeling of being, in addition
to an earthling, a spacefarer.

Lacking a view of the stars, I have subscribed to NASA
and ESA newsletters, where I read about a rainbow comet
with a heart of sponge, about the possibility of life on one
of Jupiter's moons. I follow the social media accounts of
various telescopes, I listen to podcasts about comets and
black holes, and I email with an aerospace engineer about
a future colony on Mars.

In addition to following the national weather service's
forecasts, I also watch a space-weather report by Tamitha
Skov, a meteorologist who reports on fire and plasma
storms, rather than a rainy holiday season or fog-bound
freeways. The screen behind her shows not an inland high-
pressure system but the churning surface of the sun.

Her reports are intended for sectors that are influenced by the conditions between the sun and Earth, like commercial aviation. But for me, Skov's predictions are mostly a reminder of the mind-boggling fact that Earth is part of a solar system, a planet among planets. A reminder of where and who we are. The water in which we swim.

Until recently, I had no name for what I was looking for. "An overview effect on Earth," I would say to friends who asked, and then I would get stuck halfway through explaining what that was, because it sounded like I was grasping for an earthly imitation of what a few people have experienced in orbit. I want a similar experience, but an earthbound one. The same kind of awe, with both feet firmly on the ground.

I kept fumbling through words that could best express my longing, until last week, when I read the theologian Wil van den Bercken's book *Uit sterrenstof gemaakt* (Made of stardust), and therein found the term that seemed to capture it: *cosmological awareness*. Van den Bercken describes it as "the profound awareness that we live in a, statistically speaking, negligible planet in an unfathomable universe."

WIL VAN DEN BERCKEN RECEIVES ME in his Utrecht apartment on a chilly Thursday morning with coffee and cake. I have invited myself here to find out more about the impact of cosmological consciousness on his thoughts and actions. He launches into an explanation before I've even taken my first bite of cake. This consciousness, he says, has given a different slant to day-to-day life, and he wishes his fellow earthlings could see this, too.

"If everyone began their day with what and where we

truly are. If we were continually aware of the fine-tuned alignment of natural forces and circumstances necessary for life to emerge—if the Big Bang had been a tiny bit slower, everything would have imploded; a fraction quicker and atoms, and thus matter, would not have formed. If we really appreciated the improbability of all this, then wouldn't we take better care of each other and our planet?"

I think of the absurd numbers in his book. The chance that a cosmos like ours could form is 1 in $(10^{10})^{123}$: zero-point-billions-of-zeros.

We would not exist if space were not three- but four-dimensional, if electrons did not have the exact mass they have, if the electromagnetic force were weaker and gravity stronger. If, if, if. An absurd cosmological balancing act is holding our lives together.

Van den Bercken refills the coffee. "I experience life differently since allowing myself to be awed by where we are," he says. "It puts things in perspective. Especially the political craziness. All the lines we draw between ourselves and the others are absurd in light of our place in the boundless cosmos."

It's exactly the *lightness* that comes with putting things in perspective, I realize here at this small kitchen table, that I feel stirring when I read about meteors and follow the space-weather report. Researching something bigger and far more complex than anything here on earth seems to make daily life a bit more orderly and manageable. If we can survive in an ice-cold, pitch-black cosmos, then surely we can close the gap in a polarized neighborhood, come together to fight climate change, and offer millions of refugees a new home? Then we can, in fact, do anything?

Cosmological awareness reminds us how small and lucky we are, but doesn't it also make us bigger, stronger? It doesn't show us only how inconsequential we are in the big scheme of things but also how exceptional. Van den Bercken talks of the billions of solar systems throughout the universe and I look at him: the theologian and his coffee mug versus eternity. You can either nurture awareness or neglect it. You can give it a workout, like a muscle—and like a muscle, it will atrophy if you don't exercise it. Awareness is not static, it is an ongoing process. Cosmological awareness, therefore, is a matter of focusing your attention on the cosmos. And van den Bercken has shown what that can achieve: a vision of the world that approaches the attitude of an astronaut.

IN THE WEEK AFTER MY MEETING with van den Bercken, I dive back into studies, books, photos, films. I find all sorts of things, but attention requires focus, and this is exactly what my search lacks. In the endless expanse of the internet, I open tab after tab, window after window. My head spins from the timelines, emails, and streamable lectures. Astronauts have it easy: looking at Earth from space, you don't have much choice about where to direct your attention. But looking into space from earth is another matter. My attention wanders past planets, nebulae, and supernovas without stopping at any of them. They all feel equally far away. I lack a stepping stone, a starting point.

One evening I read the picture book *Bringing Down the Moon* to my children. Mole stands on a hilly lawn and wonders at "that shiny thing" in the sky. "I'll jump up and pull it down," he says. My children laugh at the silly mole reaching upward with his burrowing paws. "You can't!" the youngest yells at the illustration in the book. On the next page, Rabbit corrects Mole. "You'll never do that. It's not as close as it looks." Eventually, Mole gives up, but later in the book he finds the moon again, this time reflected in a pool of water. Elusive, but closer than he could have imagined.

"Is the moon further away than the dino time?" the eldest asks. I nod absently; my thoughts are still with that picture book. The moon being close by and far away at the same time. I am reminded of the study of astronauts and the overview effect—distance begets proximity—and then of the psychologist Annahita Nezami and her goal of recreating that effect with VR goggles for us here on earth. I remember what she said about how to come closer to experiencing the overview effect: allow yourself to be awed by something larger than humankind. Look up at the stars. My attempts failed because I went in search of stars that—even from an observatory—were largely invisible from here.

Why did I ignore the moon? Because I take it for granted? I should have thought of this earlier! The stars might be swallowed up by artificial light, but we can see the moon fine from our square. I quickly put away the picture book, switch on the night-light, and go downstairs, through the living room, to the balcony. There it is, half-visible, right above the new high-rise hotel up the road. I resist the urge, like Mole, to reach out and touch it. I stand for a while, my head tilted back, staring at our arid satellite. As a child I once found a chalk-colored rock full of fossilized shells at a building site. "These rocks must have turned up when they were drilling," the teacher said the next day when I brought it to school. She also said it was probably millions of years old. It was my first fossil; I treasured it as a signpost to a deep past. The moon has the same pockmarked surface, and exactly the same color. A prehistoric imprint.

7

•

The Secret Breathing of Earth

ON AN AUTUMNAL SUNDAY AFTERNOON, I TAKE THE METRO into town for a lecture by the well-known British astronomer and self-proclaimed "lunatic" Maggie Aderin-Pocock. She appears on the large stage of the Zuiderkerk (a former Amsterdam church, now in use as a multipurpose public space) wearing a white tulle dress; draped over her left shoulder is a glossy bag emblazoned with a silver moon. Aderin-Pocock is on the physics and astronomy faculty at University College London; she's worked on the construction of the Gemini telescope and the Aeolus satellite, been awarded honorary doctorates from five different universities in the UK, presented *The Sky at Night*, one of the BBC's longest-running programs—and this is only about half of her résumé.

I read in an interview she gave that the moon played a decisive role in the course of her life. The daughter of Nigerian parents, in 1970s London she felt she didn't belong—

until she looked up at the pale satellite that stuck with her, night after night. The moon, she thought, belongs to everyone, including me.

At fifteen she built her own telescope. By following the moon, Aderin-Pocock arrived at where she is now. Hers is a story of someone who dared to navigate by a celestial body, who from a young age determined her path in life by looking up. If there were a course in cosmological awareness, this woman's life story would be the lesson plan.

ADERIN-POCOCK TAKES HER LISTENERS on a high-speed ride with her fascination. Let's start at the beginning, she says, about 4.4 billion years ago. The most widely accepted theory for how the moon formed is that the young Earth collided with another planet, whereon a chunk of Earth's mantle shattered and was catapulted into space. That debris slowly clumped together and became the moon.

A wayward ball of earthly debris, one quarter the diameter of the earth and with a mass of just over 1 percent of its mother planet. Since its formation, the moon has been slowly moving away from Earth, at the rate of about one and a half inches per year, which is about how fast our fingernails grow.

Aderin-Pocock keeps up her tempo, jumping five hundred million years ahead in time, when our solar system entered a period called the Late Heavy Bombardment. Like other celestial bodies, the moon was bombarded by planetoids, which left behind the craters we still see today. The moon's landscape hasn't changed much since then. The main reason for this is that the moon's atmosphere is so thin—no wind, no rain—erosion is practically nonexistent. So thin it's not even called an atmosphere, but an *exosphere*. The temperature differences are enormous: noontime temperatures at the equator can reach 212°F, and at night, negative 238°F.

My head is spinning from all this information, but I have to keep up, because Aderin-Pocock has moved on again, now to the oldest lunar legends. Around 2000 BC, one legend goes, a Chinese bureaucrat tried to shoot himself to the moon on a large chair lashed to early rockets. The Roman Syrian Lucian, in AD 150, wrote the first science-fiction story about the moon. And in the seventeenth century, the renowned mathematician and astronomer Johannes Kepler injected his scientific knowledge into a fantasy story about a witch who could transport people to the moon. (Fiction fatefully turned to fact when Kepler's own mother was accused of witchcraft.) The moon, Aderin-Pocock says, was seen as a protectress for those unwelcome on earth. The witches, the drunkards, the werewolves.

And then we're back in London, with a girl whose family calls her a "lost Nigerian." A girl who doesn't feel rooted in England and, like a modern-day seafarer, looks to the sky to understand where she is and where she must go. After the talk, I buy her book and we talk briefly about her research and my quest. When I ask if we can take a photo together, her ten-year-old daughter suddenly appears beside us, smiling enthusiastically. "Three lunatics in a row," Maggie laughs while someone from the organization takes our picture with my phone. In the metro on the way home, I look at the photo. Three females (one of them in a tutu) of different ages and colors, connected by a fascination. The top scientist alongside the anxious writer and the exuberant symbol of a new generation. What about this picture moves me so much? It's what Aderin-Pocock said while we were posing—"three lunatics"—magic words that made us, for that brief moment, equals. United in a shared destination, regardless of the paths we have taken so far. Three lunatics: bound together not by earthly identities, but by a shared view.

It's dark when I get out of the metro, and I see, above the rooftops, the moon rise. A clump of earthly debris that's slowly leaving us. It's a sad idea, a fellow traveler going off on its own. I wonder what all that debris would have become if not for that collision, if it had just stayed part of Earth. A forest floor, the tundra, my front yard?

FROM OUR BALCONY, I start looking at the moon—I mean, *really* looking. Watching it wax and wane, noticing its changing hues. I begin to feel the strength of its light, even in our overlit city. Sometimes I see the "chin of gold" that Emily Dickinson describes, sometimes I see what Sylvia Plath means by "white as a knuckle and terribly upset."

There are evenings when the moon suddenly moves me, its pockmarked face hanging above our hodgepodge of a neighborhood, its soft glow of secondhand sunlight. It's a comforting thought that something is looking down on us, even though it's a dusty, lifeless eye.

One day, David comes home with a telescope and a map of the moon. Together we read the names: Sea of Clouds. Marsh of Sleep. Lake of Forgetfulness.

These names, of course, contradict what we now know to be an arid surface. They mainly reflect what we once thought we saw. It is hard to observe (and name) the unknown. You look for points of reference, project the familiar onto the alien. You fill the mysterious plains and craters with water.

The first time I saw the moon through the lens of the telescope, I was taken aback by its sudden closeness, its gray-white surface so incredibly sharp. For a moment I'm overcome by the ridiculous thought that I'm doing something illegal, that I'm trespassing. As though humans can't travel such distances with impunity.

There is another strange sensation: the absence of sound. I know that we can't hear anything on the moon—sound waves need air to travel—but now that it feels so near, I almost expect to hear something. The wind. A welcoming voice.

The eighteenth-century Chinese astronomer Wang Zhenyi (a woman!) wrote that, although the moon might make no sound, "something at its center may enlighten its listeners." Sometimes, looking is a kind of listening.

I seem to have found my way to the Sea of Tranquility, where Neil Armstrong took that elegant first leap onto the dusty surface. That first moon landing, the Apollo 11 mission, celebrated its fiftieth anniversary in 2019. That unlikely journey was on everyone's lips. The courage it took to travel those 240,000 miles. Documentaries and articles came out, profiling those three men in their *Saturn V* rocket, the music they played, the instruments at NASA's Mission Control Center—but strangely enough, almost no one talked about the moon itself.

That mysterious landscape was suddenly within reach. The highlands and the "cold traps," areas where the temperature is so permanently low that volatile chemical compounds are accidentally trapped there for good, because they cannot warm up.

No one talked about the far side of the moon (which for us is still unseeable), over which the Apollo astronaut Michael Collins flew, solo, in the command module while his colleagues Armstrong and Buzz Aldrin stood on the moon's surface.

The far side of the moon. Collins was completely on his own, unreachable by mission control in Houston. An experience on which he later looked back, concluding that no man since Adam has ever been so alone.

No one talked about the moon, which has such an effect on Earth's surface. The reason for the tides and the mysterious "tidal bore" that can occur in rivers during major ebbs and floods. There's also something called "terrestrial tides," I read in the book *La Lune est un roman* (Once upon a moon) by the French astrophysicist Fatoumata Kebe. She describes how the solid surface of Earth, reacting to the pull of the moon, rises and falls about a foot or more per day.

Kebe calls it "the ground's secret breathing," invisible to the human eye because the entire surface rises and falls over an area of hundreds of miles. A secret breathing, she writes, is part of the "universal sympathy," a term used by the Stoic philosopher and astronomer Posidonius to explain the tides. "Sympathy," Kebe writes, "is what we feel together, with another person, with nature, with the entire universe. Sympathy exists between beings and things."

Sympathy. Derived from the Greek, meaning "feeling together, experiencing together." Being in sync. Without this synchrony between Earth and the moon, our planet would be thrown off-kilter.

THE EVENING THAT I FINISH READING KEBE'S BOOK, the waxing half-moon hangs in a cloudless sky. Even without the telescope, it feels nearby, no longer a strange, distant rock but an element of the cosmic movement that is us. The astronauts who looked back on our planet from space were in awe of the connectedness of life on earth; looking from Earth to the moon, I start to realize that this connectedness does not stop at our atmosphere. The black night is not a void, but an expanse of forces that sets us in motion, lifts us more than a foot in the air every day.

I look out at the basketball court on the square in front of the house, where four boys bounce the ball across the asphalt. I imagine the court's surface undulating beneath their sneakers. I recall Wil van den Bercken summing up cosmological awareness as "the profound awareness that we live in a, statistically speaking, negligible planet in an unfathomable universe."

But maybe you could also put it this way: the awareness that everything up close is connected to something far away, that the known is set in motion by the unknown. Even what looks solid, like the ground beneath our feet, is in constant conversation with something in the distance, and secretly breathing.

8

·

An Answer to the Distance

ONE EVENING, MY NECK IN PAIN FROM HUNCHING UNDER the telescope, I read the poems of the Canadian astronomer Rebecca Elson, a specialist in black holes and the life cycle of stars.

> We are survivors of immeasurable events,
> Flung upon some reach of land,
> Small, wet miracles without instructions,
> Only the imperative of change.

I stumbled upon her work while researching the Hubble telescope, which took the picture that began my journey: the Hubble Ultra Deep Field, the shards of light.

Elson started writing poetry while studying Hubble data. Maybe, at the sight of so much mystery, science wasn't enough. "Science describes accurately from outside, poetry describes accurately from inside," wrote the science-fiction author Ursula K. Le Guin.

I cannot ask Elson what made her turn to poetry, as she died in 1999 at the age of thirty-nine, the same age I am now. But I hope that it had to do with the Hubble telescope. That the endless view goes both ways. That she wanted to understand the entirety, from without *and* from within.

In one of her poems, she calls upon her fellow scientists not to underestimate poetry. "Because curiosity, after all, is also of the spirit." Elson writes what I want to read: that a poet like me can grasp the reality of the universe just as well as a scientist. Data, explanations, and facts, she says, are different from understanding. Understanding comes from paying attention to the world around us, a kind of attention that requires one to be faithful to the soul. "Facts are only as interesting as the possibilities they open up to the imagination." I had always interpreted the fence separating the European Space Agency's museum from its laboratories, culture from science, as a line between those who know and those who don't. But if it stands for anything, I think now, then it's for the hard border between facts and imagination. Elson tears down that fence.

In her poem "We Astronomers" she describes what she calls a "responsibility to awe." Our duty to the awe-inspiring universe. She exercises this obligation herself through her tireless expression of her enthusiasm for outer space. For the first time, I hear it as an answer: "response-ability."

For Elson, the cosmos doesn't only awe us, it awakens something in us. The yearning to respond to that awe. And that is exactly what I feel as I make my way back to the telescope and scan the moon with my left eye. The yearning to formulate a response. As though that white thing in the distance is asking me a question. I've no idea *what* question; maybe it's not even a real question, but a kind of space only a question can create.

I slowly pan across the bays and meandering grooves, whose origins are still a mystery. Looking through a telescope is precise work: the tiniest shift down here equals a leap of thousands of miles up there. Sometimes I try to get my children to look, but at the first glimpse of the moon they jump up and down with excitement, and we lose sight of the whole thing. So I try to wait until after the kids are in bed.

IN NASA'S NEWSLETTER I read about preparations for a crewed mission to the south pole of the moon. At the same time, NASA and the European Space Agency are jointly working on a permanent space station orbiting the moon, from which missions to the lunar surface will be easier to launch. I read about Israel's and India's recent, unsuccessful uncrewed lunar missions, and about the private Japanese-Luxembourgian lunar robotic-exploration company, ispace, which plans to mine for minerals on the moon's north pole.

My computer's desktop is filling up with purring PR videos promoting similar lunar mining companies, utopian websites about future moon colonies, and photos of gleaming rockets. All of them with the same message: we have to go to the moon, and quickly. Each has its own reason. There is the political vanity of national governments, which see planting a flag on the moon as a show of extraterrestrial power. There are companies that want to mine hydrogen and oxygen for rockets stopping to refuel on their way to Mars. There are the scientists who regard the moon as Earth's ear: radio waves from the cosmos are more readily detected on the far side of the moon (the side we never see), protected as it is from all the radiation and white noise we generate here on Earth.

THE MORE I STUDY that empty landscape above us, the more I realize how vulnerable that emptiness is. The moon is no longer just a natural phenomenon, it is now also political theater. "Wherever you go, there you are," said the legendary mindfulness guru Jon Kabat-Zinn. So everything we are here, we are in space, too. And now that it is within reach, the moon cannot evade our greedy clutches.

Fair enough. And yet . . . shouldn't space be for everyone? Isn't that inherent in the word? *Space.* "Continuous, unbounded, or unlimited extension in every direction," says the *Oxford English Dictionary.* And: "The immense expanse of the universe beyond Earth's atmosphere."

An organization calling itself a fellowship of Persian storytellers is said to have written NASA before the launch of *Apollo 11*, begging them to abandon the mission. A moon landing would rob the world of its illusions. Once you land somewhere, they argued, then you no longer need to yearn for it.

For the first time, I wonder who the moon actually does belong to, and I discover that there is such a thing as "space law" (a more robust term is hard to imagine), a body of international law governing space-related activities. But deeper analysis shows that space law is not so robust after all. The agreements and treaties are based, more or less, on the Outer Space Treaty drawn up between the United States and the Soviet Union in 1967. This pact dates from the Cold War, when the space race was a competition between two political superpowers. It did not foresee the commercial exploitation of the universe, and the issues this would entail.

In an episode of *Tegenlicht* (a Dutch investigative journalism program focusing on politics, economy, society, and science), space jurist Frans von der Dunk explains that there is a legal vacuum in space. He is concerned about the plans for mining and tourism, potential military intentions, and the millions of kilograms of waste that are already floating around Earth. Making rules about space garbage is one thing—everyone is aware that it has to be cleaned up, otherwise one day there won't be a safe path into space. But keeping space demilitarized is another story. At present, it is forbidden to carry out military activities on the moon, and countries may seek to verify one another's compliance, but the law's wording is cloaked in generalities. "Back in 1967," says von der Dunk, "no one saw much point in specifics, but now that time has come. There have to be clear rules regarding compliance verification. A certain trust must be established." That is a tall order, for more than half the world is, in one way or another, active in space.

He argues that an update to the Outer Space Treaty is long overdue, and considering the speed of developments in this arena, it is a matter of some urgency. The moon, together with international waters, belongs to "mankind's common heritage."

This term strikes me as strange, because who exactly did we inherit all these unreachable places from? Nevertheless, I'm glad to know there are wildernesses of water and dust that have—for now—evaded our appetite to gobble up everything in sight. For now. Who knows, maybe in a few years the moon will be a multifunctional refueling and radio station, and we'll be surrounded not by space but by a thick layer of garbage.

An animation at the Space Geodesy and Navigation Laboratory in London shows the increase in space debris between 1957 and 2015. For the first few seconds, Earth floats like a clear, colorful marble in the endless, black sea. And then, gradually, the blackness fills up with small white dots—disused satellites, paint chips, bolts, lids, jettisoned rocket parts. By 1970, Earth's atmosphere already appears to be clogged with waste, but then we have another forty-five years to go. In 2015, the marble is completely invisible, enveloped in white dots—twenty thousand pieces of space rubbish—like a layer of mold. Twenty thousand bits of junk, reminders of humankind's craving for exploration. And that was back in 2015.

I feel stupid as I close the animation. I've always regarded space travel as something romantic. Humankind traveling through an expanse of mystery—it doesn't get more poetic than that. As the self-anointed house poet of the European Space Agency, I was inspired by replicas of the *Apollo* lunar module and the *Ariane* rocket. I saw the universe as another world, or rather, a nonworld. Alien and mythical, the way the astronomer Carl Sagan is said to have put it: "Somewhere, something incredible is waiting to be known." And I associated space travel only with that yearning for the unknown, not with floating junk and pollution. How does space congestion affect our potential to be awed, and to zoom out on ourselves? I think back on the cynical observatory attendant and her comment about intergenerational amnesia. What will the moon mean to my children? To their children?

ONE DAY I come across the Moon Village Association (MVA), a nonprofit organization whose goal is to ensure that the moon continues to belong to everyone and, at the same time, to no one. It sounds like something from a children's book, a fairy-tale organization that attends to outer-space business. Curious and a bit skeptical, I call the founder of the MVA, Giuseppe Reibaldi, an Italian engineer and longtime employee of the European Space Agency.

Reibaldi is a fast, enthusiastic talker, his sentences tripping over one another. He is of the opinion that we have too easily turned over the moon to big business. Our laissez-faire attitude allows issues like mining and colonization to go undebated, giving free rein to companies and organizations already active in space. "We have to do something about this moon madness!" he exclaims. Even if—*especially* if—you don't have any technical know-how.

Reibaldi's love affair with the moon goes way back. When he was nine, he started writing a science-fiction story about a gang of villains who go to the moon and are attacked by monsters. He never finished the book, but you could say he was doing so now with his Moon Village Association. A final chapter where the moon itself brings together all earthlings as brethren. For this is essentially the goal of his nonprofit. Reibaldi uses other words for it, but what he is hoping to do is use space to bridge divides on earth.

His waterfall of words cheers me up. It's not just his melodious Italian accent, it's also the fascination his story exudes. I am always happy to talk to people like Reibaldi, Annahita Nezami, and Wil van den Bercken. People who have been at it for some time and who reassure me that I'm not just tilting at windmills or that space isn't an illogical detour to take when you're looking for a new perspective on earth.

"Seen from the moon, we are all the same size," the famous Dutch author Multatuli wrote in 1879. Reibaldi echoes this thought, with a slight twist: looking at the moon reminds us that we all belong, or don't belong, equally to this corner of the solar system.

An actual "moon village," Reibaldi says, is still a long ways off. A permanent settlement on the moon is not only prohibitively expensive but dangerous as well because of long-term exposure to cosmic radiation.

So for now, the Moon Village Association uses the moon as an international hypothetical project. A way of engaging our collective imagination. "We all have a stake in the moon," Reibaldi says. We must put our heads together as to how our actual presence up there will look; otherwise the multinationals will decide that for us. What we need is a fully shared dream, because regardless of what NASA said, the Apollo moon missions were far from a global dream for all humankind. The crew and the control room were too white, too male, too American for that.

THE MOON VILLAGE ASSOCIATION has 220 members worldwide, plus 26 member organizations, including planetariums, universities, and engineering firms. They meet weekly to discuss all possible aspects of a human community on the moon. Ethics and imagination, technology and opportunities.

I ask him how I can participate. The next Moon Village Association meeting is a week from now in Budapest, but Reibaldi tells me I don't have to go. "Start right where you are. Bring people together and open a discussion about the moon. It feels strange at first, of course, but chances are you'll be pleasantly surprised. Most people, regardless of what they do or who they are, look up at the moon and are curious about it. And most people like sharing their fantasies and dreams about space. Looking at the night sky and finding meaning in it is something mankind has been doing since its very beginning."

I am reminded of Rebecca Elson's poem about our responsibility to awe. For her, though, talking about space was completely natural: aside from being a poet, she was a brilliant astronomer, on the faculty of universities like Harvard and Cambridge. How am I supposed to start a discussion like this without people thinking I have a screw loose? A confused woman with cosmic delusions: "Hello, how do you see our future on the moon?"

Reibaldi understands my hesitation, but he has a solution. "The Moon Village Association is drawing up a questionnaire to find out how the average person relates to our closest neighbor. Our plan is to conduct the survey worldwide, but we're still working on the content."

Maybe I could test out the questions, he suggests. Then I can not only send feedback to the Moon Village Association but also have a conversation starter of my own. At first I think he's joking, that soon he'll suggest meeting at

Platform 9¾ and shooting to the moon in a big glass ele-
vator. But when I realize he's serious, I get serious, too.
Why not? People spend all day talking unashamedly about
colds, about housing prices, and about misunderstandings.
Why not talk about our common view?

9

•

Museum of the Moon

IT IS A MILD, VERY WET AUTUMN. WHILE HUGE WILDFIRES rage across America and Australia, heavy rainfall has flooded our street. A strong westerly wind pushes the water through the gaps in the window frames. The handyman the owners' association sent is unable to do anything about the leaking. These houses, he says, are not made to withstand this kind of weather. He tells us they plan to put some sort of transparent waterproof coating on the outside of the houses in a couple of years. The idea gives me the creeps, our house in an impermeable membrane, like a trapped embryo.

Giuseppe Reibaldi sent me the draft questionnaire last week, and I've resolved to head out to the square today with the Moon Village Association's questions. Start right where you are, Reibaldi said. It sounded so logical, but now that the moment has come, I have my doubts.

I see myself standing there holding a stack of printouts,

like the Jehovah's Witnesses handing out their paradise-touting leaflets at the ferry pier.

Back when we had just moved here, I might have dared. But that was before I caught on to the fact that I, with my mortgage and my fair-trade sneakers, represent exactly what is wrong with our neighborhood. Before everything around me began to change to my advantage. A café with vegan sandwiches, a sustainably minded clothing store, an art-house cinema. The scary part was that I didn't have to do anything for it. Simply bring together enough well-educated newcomers and it seems to happen by itself, at the expense of what was here before. It is a strange paradox: doing my best to fit in has only widened the gap between my neighbors and me. The more visible I am, the worse it gets, so I try to keep a low profile, resist giving in to my tendency to always try to make things better, to get things moving. Pushing a survey about our common future on the moon doesn't exactly fit in with this strategy.

Besides, I wonder if there's anywhere I would be able to shake off my inhibitions. I need a context where I won't come across as a missionary or a nutcase. I look from the questionnaire to the clouds above the square, and suddenly it comes to me: what seems crazy in daily life—a woman with a spiel about the moon—can find acceptance if it's an art project. What I need is an artistic cover.

THREE DAYS LATER, I'm standing in Lille-Flandres train station, just over the Belgian-French border. A gigantic moon hangs from the concourse ceiling. Held in place by cables and wires, it floats a few yards above the incoming and outgoing passengers' heads. It is bluer than I am used to; all its craters, mountain ranges, and seas are drawn to scale. The installation, called *Museum of the Moon*, has been traveling around the world for two years. A contemplation of our shared view, I read on the project's website.

In the station café I take a bite of dry croissant, wash it down with bitter coffee. I got up way too early, and the fatigue is giving me an insatiable appetite. Lille is not that far from Amsterdam as the crow flies, but what with all the transfers, it took me five hours to get here. A far-too-complicated train route crossing two national borders. I arrived at 12:30, and if I want to get back home today I'll have to leave by mid-afternoon. Before I left, David told me I was insane. "You're schlepping all the way to the north of France for *what?*"

But I knew immediately when I saw a picture of the installation, the artificial moon that turns a station into a museum. I decided to go see this northern French *lune* and make myself part of the project. An ad hoc museum guard with a few questions about the celestial object dangling above our heads.

Perfect plan, in theory. I hadn't counted on the travelers' haste. The *Museum of the Moon* does not appear to be anyone's final destination, everyone is passing through, they stare anxiously at their cell phone or the departures board, drag luggage and/or children. No one stands still for even a second. Even here in the café, everyone's in a hurry. I don't dare strike up a conversation with stressed-out commuters chugging down coffee from plastic cups.

AN HOUR LATER, I'm still sitting in the café. I stare at the travelers who flow like the tides toward the platforms while the solitary moon hangs above them.

I am struck by the overwhelming presence of military police. Armed men in every corner. The moon is supposed to turn this space into a museum, and these guys are turning it into a war zone. How long before the first weapons land on the moon?

This reminds me of the six American flags on the moon that Milo Grootjen, house astronomer at Amsterdam's Artis Zoo, told me about. These flags were planted by Apollo crews, but by now have been bleached by the sun. Six white "truce" flags on the moon.

Every once in a while, the odd passerby actually stops to look. But every time I reach for my coat, ready to approach them, the moment has passed and the person walks on.

I trudge to the bar to order another bitter coffee. Back on my barstool, I reread the multiple-choice questions for the umpteenth time.

In the context of climate and environment issues on Earth, would you say that developing long-term infrastructures on the Moon:

A. is a waste of money, considering our needs on Earth
B. is an opportunity to exploit lunar resources to relieve energy shortages and other natural resource issues on Earth
C. is an opportunity to test alternative energy production and waste management methods, offering possible solutions for Earth
D. Other:

How would I respond? My first answer would be A. And

B. And C. And, yes, D, "other." Maybe it's too early for lunar infrastructures. Before we start building, we need to establish the moon's actual significance for society, for humanity. Not only economic and scientific, but cultural and spiritual as well. The word *infrastructure* makes me think of the "canali" on Mars. The dark, straight lines on older surveys of the Red Planet that the Italian astronomer Giovanni Schiaparelli believed to be waterways: structures that would indicate intelligent life. MARTIANS BUILD TWO IMMENSE CANALS IN TWO YEARS: VAST ENGINEERING WORKS ACCOMPLISHED IN AN INCREDIBLY SHORT TIME BY OUR PLANETARY NEIGHBORS, a *New York Times* headline blared in 1911. And the Portland *Oregonian*: MARTIANS FINISH CANAL ON PLANET.

It was most likely just an optical illusion, akin to seeing seas on the moon. The human tendency to see patterns that aren't there. The penchant for weaving disparate elements into a narrative.

HIGH ABOVE THE COMMUTERS, the artificial moon seems to rock slightly. Is there a draft, or am I just tired? On my phone I read an interview with Heino Falcke, the physicist who led the team of scientists that in April 2019 succeeded in photographing a black hole for the first time. Falcke talks about the limits of the observable universe and his belief in God. There are, he says, many parallels between science and faith: both seek to understand the essence of things. Eternal space, for him, is a metaphor for what he calls God.

The interview runs in the science section of the newspaper, even though it is mostly about theology. I'm starting to notice how one-sided commentary on space is. News about space is sometimes front-page news, sometimes it is an international item, but mostly, like now, it's in the science section. But space travel is also culture, politics, philosophy.

In the prologue to her book *The Human Condition*, the philosopher Hannah Arendt invokes the launch of *Sputnik* in 1957 as an example of the complexity of the modern world. We leave important things to the scientists, engineers, and politicians, because these things have become too complex for us "regular" folks. Whether humanity will grow as it conquers space depends, says Arendt, on how connected with space we are in our stories, in our language. If we become further estranged from space because of high-tech science, then there's no benefit, only loss. Issues that fall outside public debate, she says, are out of our reach. What would Hannah Arendt say if she saw me here in this train station? Is that what she meant when she wrote that space has to be kept within our reach? Would she slow down under this moon?

A WOMAN HOLDING A LITTLE GIRL BY THE HAND, their black locks pulled into identical ponytails off to one side, stops and looks up. The woman points, the girl's gaze follows her index finger. I slide back my stool, throw the questionnaire into my bag, and hurry out onto the concourse.

But they've already moved on. My phone vibrates. A text message. *Who's picking up the kids?* I could say it forty-two times, and still he doesn't remember. *Home late*, I text back. I walk over to the kiosk and check out the pasty-looking sandwiches in the cooler. Why didn't I bring food from home? Why didn't I *stay* home? I look at my phone: the next train to Amsterdam leaves in forty-five minutes.

Now I see myself standing there on the concourse. A tired woman with bags under her eyes and wearing a too-heavy sweater—it was still freezing cold on the ferry this morning—snot stains on the shoulders, hair a mess, searching eyes. A woman who not so long ago breezed through life. Until, little by little, she saw cracks in everything. The brokenness.

No, I always saw it, but I had found a way to live with it. But not anymore. The chasm in my own gentrified neighborhood, the devastating consumerism that I just coast along with, the current hard-right tilt in the Dutch social and political landscape, the drought, the feeling that everything that's supposed to stay together is coming apart at the seams. I am not going to fix this by bugging random commuters in a train station in Lille with some questionnaire. And yet, it feels like I'm on the right track. Because up there, hanging from the concourse ceiling, is a reminder of our common view. A reminder of what we share, *that* we share.

Sometimes I'm so afraid that we've become so good at announcing our differences, in defining ourselves, our identities, down to the last detail, that there is simply no more room, or language, to talk about what connects us. Words die out, too, and with them, ways of thinking. Of course things have to be hammered out, painful fault lines must be called out, but what if all we see are the cracks and we've run out of glue? If we jettison the metaphors that relate to all of humankind, no longer realize that we form a unique group? A group formed by materials from dead stars somewhere in the universe. A group that you won't find anywhere else, as far as you look.

I sit on a bench next to the kiosk, take a bite of a gummy sandwich, and look around me. In the middle of the concourse, a man in a red jacket slows down. His gaze turns upward. I quickly get up. But before I can take a step, a gruff military policeman directs me outside. A suspicious package has been spotted. The concourse must be cleared. We're pushed outside onto the forecourt, the station doors shut, and stay shut.

STRANDED TRAVELERS MAKE CALLS and stare anxiously at the screens on their smartphones, each of which has a stronger processor than the computer on board *Apollo 11*. Farther up, young men are made to empty their bags, mothers rock whining children, and elderly people look for a spot on one of the scarce benches. A group of irritated backpackers sits on the traffic barriers alongside the bicycle path. Policemen stomp aggressively across the forecourt. Everyone in their own story.

A railway employee instructs us through a megaphone to walk to the next station and continue our journey from there. The crowd plods off, the forecourt empties. Mission unaccomplished.

I follow the procession, knowing the survey is a lost cause now. In a situation like this, every form of abnormal behavior is suspicious. Soon I'll be the woman who created a diversion by jabbering about the moon when a bomb went off at Lille-Flandres.

My sense of failure grows with every hurried step through the streets of Lille. Our pace, I realize, makes it impossible to strike up a normal-sounding conversation. We are people in a hurry, people who are not where we should be, people already mentally on the next train. I scan the crowd for a face not looking straight ahead, I try to slow down, but my legs won't cooperate. It's strange, how contagious anxiety is, and how it closes in on us, knotting us up in a stifling here and now. And robs us of vistas and vision.

10

•

Sunset on Mars

ONE WARM WINTER AFTERNOON, I SIT IN THE PALLID SUN-
shine outside a recently opened café on our square, eating
a green pea hummus sandwich and listening to a podcast
called *The Habitat*. It's a series about a NASA dome in a
remote area in Hawaii where six selected people—doctors
and scientists—spend a year living as they would on Mars.
The three men and three women, under constant monitor-
ing, were part of a long-term study of the riskiest factor
of a crewed mission to the Red Planet: the human psyche.
Similar experiments have been conducted over the years
in a variety of locales and by various space organizations,
but this time the podcast producers gave the participants
a device with which to record their own experiences. The
study revealed their day-to-day life to be carrying out routine chores, filling in questionnaires about their physical
and mental well-being, and quibbling about the commu-
nal breakfast.

Two of the crew members become romantically involved, irritation arises over a didgeridoo someone brought with them, there is gossip and bickering. But despite all the minor annoyances, I am impressed by how easily these extremely intelligent adults surrender to this yearlong fiction.

They are not on Mars. They can leave their dome at any time without peril. And yet they stick with it, that endless fussing with spacesuits in a decompression chamber, as though their lives really depend on its painstaking execution. They are on Mars on Earth, and I am touched by their dedication to a fantasy, a deal, a dream.

After my botched moon trip to Lille, I've turned my attention to our neighboring planet. I want to enlarge my circle, break out of "cis-lunar space," the region from Earth to just past the moon. Space that, after all I've read and seen, hardly feels like "space" anymore, what with all the satellites and scrap metal. Space that, now that we're there, we no longer yearn for the way we used to—exactly what the Persian storytellers warned us about back in 1969.

I decide to zoom out even farther, past where the astronauts went, to a place no one has been, where I can mentally break free of Earth because no human boundaries have been drawn yet. A place you can still dream about and yearn for. Deep space.

MY FIRST DEEP-SPACE DESTINATION IS MARS. Traveling to Mars—as yet impossible for humans—will be a completely different story than going to the moon. Seen from the fourth planet from the sun, Earth is no longer a shiny blue marble, but a vague dot. Depending on where they are in their orbits, the distance between Mars and Earth is anywhere between 34 and 249 million miles. From that distance, there's no looking back.

In an ESA newsletter I read about MELiSSA, the European Space Agency's longest-running project, begun in 1989. The name is an acronym for Micro-Ecological Life Support System Alternative. In the MELiSSA Pilot Plant in Barcelona they are working on a circle, or rather a loop, of life in which algae, plants, and humans (or, so far, rats) sustain one another. A compact recycling system where organic waste is converted into valuable raw materials, enabling astronauts to be self-sufficient during the monthslong journey to Mars. MELiSSA seems to me the best place to expand my cosmic awareness with a mental trip into deep space.

So from that café terrace I send an email to the lab in Barcelona, asking if I can come for a tour. While the waitress clears away my plate and I order a ginger tea, some benches farther up fill up with a group of cigarette-smoking neighborhood men, regulars who congregate here during the day. Come evening, they move fifty yards down the street to the teahouse, where loud music mixes with laughter and arguing until deep into the night.

The distance between "their" benches and "our" terrace populated by laptopping thirtysomethings is maybe six feet at most, but it may as well be light-years. One of the men runs his hand gently over the branches of a summer lilac some neighbors and I recently planted. A subsidy from our local council allowed us to bring in two truckloads of greenery, which, under the supervision of a professional gardener, we planted around the trees and the sandbox. The man caresses the plant as though he were stroking a house pet.

I look at the five men; including me, we are exactly the right group size, psychologically speaking, for a long-distance space mission. Six people. Various test missions (like the Habitat) in which quasi-astronauts live in close quarters for months at a time have shown that six is the ideal number for long-term isolation.

In addition, the researchers came to the obvious conclusion that the crew members had to be complementary in character. But how to create this harmony is not yet clear. Cultural differences can either drive a wedge among group members or bring them closer. The best example is perhaps the experience of the American astronaut who was on the International Space Station at the time of the 9/11 attacks. In his letters to NASA following the event, Earth itself played a surprisingly minor role. What he did describe was the support he got from his Russian crewmates. How they showed him how to make borscht and taught him the Russian word for *empathy*. The kindness that helped him cope with those confusing times.

Kindness. That was the same answer the director of the Space Expo gave me when I asked him if astronauts have a common denominator. "When you're squashed together into a few square meters, the most important thing is to be kind to one another." To be able to be kind, you first have to bridge the distance between you, appreciate that you're on the same mission.

I send off the email to Barcelona and pay my tab; the day-care center closes in half an hour. The man is still caressing the lilac. In passing, I smile at him—too briefly; he doesn't notice.

BACK FROM DAY CARE, my sons head straight to Bob's front door. They want to see his tropical fish. Before I can intervene, the door swings open. "I saw you all standing there," says Bob. "Come on in." As I follow him into the hallway, I catch myself being surprised, yet again, that he is over eighty. He comes across as someone in his sixties. He chalks it up to all the hard luck he's had in life. Poverty and hard knocks keep you young, he says. They teach you to distinguish sense from nonsense. On top of it, he swears by a Tylenol per day and plenty of gardening.

From Bob's big leather sofa I watch the boys press their noses against the side of the aquarium. The TV is on, a game show blares through the room. Bob asks if I've had a good day, and I try to come up with a way to involve him in my search without sounding like a madwoman. I silently admonish myself as he pours me a glass of orange juice. Why make it so complicated? What is so strange about wanting to find a cohesive story from what's now a bunch of loose shards? Just start!

"Well . . . ," I say. Bob gives me a questioning look. But then the youngest eats a handful of fish food and the older one licks the wall of the aquarium. Time to go.

Once the children have been fed and are installed on the sofa for their daily portion of *Dinotrux* and David is playing chess on his cell phone, it's time for my latest addiction. I open my laptop and look up photos of Martian sunsets. The images fascinate me no end: the strange blue haze surrounding the last rays of sun, the emptiness of the sky, the ghostly nightfall. I click obsessively from one panorama shot to the next. On some of the pictures, I can see the robotic arm of the *InSight* lander, which has been studying the interior composition of Mars since 2018. It looks like it's reaching into space. Homesick, you'd almost think, for Earth.

Even stranger than the cool-blue color of the sunset is the fact that so far away, the sun even rises and sets at all. I apparently harbor the silly but deep-seated notion that the sun belongs to *us*, that it shines for *us* alone. I find it hard to imagine that we share its warmth with other planets.

But there it is, a small, faraway star in an alien sky. Under the photos taken by the *InSight* lander is a quote from the mission's director: "With many of our primary imaging tasks complete, we decided to capture the sunrise and sunset as seen from another world."

So the images aren't just collateral data from important scientific observations. No, once their assigned tasks had been carried out, mission control at the Jet Propulsion Laboratory in California instructed the robot to face the horizon, the way we might do at the end of a long workday. Capture the sunrise. So human, so romantic. The longer I look, the less surprised I would be if I spotted a person in one of those photos. It's because of that landscape, the similarities with Earth. The dunes, the gullies, the trenches that point to the possibility—once—of water.

Looking up from my laptop, I notice how unnaturally green the trees look in the light of the streetlamps outside. Then I remember that the gardening expert advised us to water our new neighborhood plants generously at first. It's been dry for days.

I put on a new episode of *Dinotrux* and borrow two large watering cans from Bob and fill them at the water tap next to the playground.

THE GROUP OF MEN is still occupying the benches. The plant petter is the first one to notice me as I walk past with the first full watering can. "Need help?" he asks. "Yes, thanks." He goes to fetch the second can at the tap, and when he's back he says, "Nice, all that greenery." The six of us spend the next fifteen minutes filling and watering.

When we're finished, we form an impromptu circle around the buckets and watering cans. A six-person crew with a simple mission: water the plants. The petter, this evening's mission leader, nods solemnly, and I nod back. Mission accomplished.

As I walk home, I have the feeling that gravity is pulling on me just a bit less. The attitude of an astronaut. A few hours later, when everyone is in bed, I receive an invitation from Barcelona to visit the MELiSSA Pilot Plant.

11

•

Beam Me Up, Spirulina

IN THE HALLWAY OF THE MELISSA PILOT PLANT I STARE AT glowing green matter in large glass cylinders. It looks like some alien substance, but turns out to be ordinary spirulina, popular nowadays as a nutritional supplement, something I normally associate with health food stores. This microalga is an excellent space traveler, four hundred times more resistant than humans to cosmic radiation. Most important, it has an ability that's crucial on a long-distance mission: it can convert its own waste products into oxygen and nutrients. Whether we ever make it to Mars depends to some extent on our partnership with this prehistoric organism.

The Pilot Plant is located in a building at the Universitat Autònoma de Barcelona, home to the MELiSSA project for the past twenty-plus years. Beside me is Francesc Gòdia Casablancas, head of the faculty of applied chemistry, biology, and environmental sciences. Only a few minutes into his explanation of the space-traveling algae and my head is already spinning from all the nitrates and nitrites. Try as I might to retrieve something from my high school chemistry class, nothing comes to mind. I find a foothold in the display across from us illustrating the basic model of the "MELiSSA loop." Six colored circles, each representing one stage in the cycle, arranged in an interlocking diagram.

The chart and Gòdia Casablancas explain how the astronauts' organic waste and carbon dioxide are recycled into oxygen and nutrients. I take notes as fast as I can. First, organic waste is sent to a bacteria mash that turns it into organic acids. Bacteria then convert the ammonia that's been released into nitrate, an important component for photosynthesis in microalgae and plants. The spirulina then photosynthesizes the carbon dioxide exhaled by the astronauts into oxygen gas while at the same time producing an edible biomass. And so we return to the humans, for whom food and oxygen are necessary for survival.

ONE THING IS IMMEDIATELY CLEAR: if I thought Mars could help me mentally remove myself from Earth, I was mistaken. Here on earth, it's easy to see ourselves as autonomous beings. Naturally, we have to eat, drink, and breathe, but we no longer regard those basic elements that sustain us as an integral part of who we are. But if we want to go to Mars, we have to face facts. We cannot leave here without taking at least some of our earthly environment with us. The journey alone takes eight months. Then we have to stay there awhile, and travel back. Transporting provisions for such a long period is out of the question, and besides, the time frame is unreliable. Self-sufficiency is far safer and cheaper. The spirulina is just as essential for such a long-distance mission as a beating heart.

Now I understand what Gòdia Casablancas meant when he came to pick me up at my hotel early this morning. On the way to the laboratory, we had to dodge fallen trees and scattered branches, remnants of a huge storm that hit Catalonia the week before. Storms are normal at this time of year, Gòdia Casablancas said, but the ferocity of this one alarmed him. Towns and cities had to be evacuated; there were fatalities. When I asked him if working on a possible escape route from this doomed planet put his mind at ease, he shook his head. "At the MELiSSA Pilot Plant we're actually doing exactly the opposite. We're not working on an escape route, we're working on connectedness. If MELiSSA is leading us away from anywhere, then it's from the myth of human autonomy."

Gòdia Casablancas is a small man with dark eyes that light up when he starts talking about salts or photosynthesis. It was he who brought MELiSSA to Spain more than twenty years ago, when the European Space Agency was looking for a suitable research partner. He proudly tells me that his team, at long last, is now nearly ready for the next big step. "We'll soon have mastered the cycle of alga- and bacteria-aided nitrification and photosynthesis. Then we can integrate plants into the cycle."

So plants, too, will be an integral part of a crewed long-distance mission, because algae photosynthesis alone cannot provide sufficient nutrients and oxygen. Gòdia Casablancas's team has set its sights initially on lettuce, wheat, and beets. Eventually the assortment will expand to some twenty species of plants, which will form a tight symbiosis with humans and spirulina. What here on earth are the ingredients for a vegan side dish will be literally our life partners in space.

Last summer I read in the American biologist and philosopher Donna J. Haraway's book *When Species Meet* that the human cells in our bodies are outnumbered by bacteria, molds, and other single-celled species. Some of these are critical to our survival, others are simply going along for the ride. Countless tiny creatures keep us alive, and vice versa. "To be one is always to become with many," Haraway writes. In this laboratory it's clear how diffuse the outlines of a living person can be.

When I share my ideas on this interwovenness with Gòdia Casablancas, he pauses for a moment to think. "That is true, but let's not forget that humans are central to the MELiSSA project. Everything we do is for the benefit of the astronauts. If you want them to be able to travel safely, you can't just construct an ecosystem; it has to be completely in your control."

Control is the main difference between nature and the MELiSSA cycle. On the way to Mars, nothing can be left to chance. Maybe you can't even call what's being developed here a cycle. There is a beginning and an end, and a center of gravity around which everything revolves: the human being who must stay alive, no matter the cost.

You could say they are actually working on the opposite of genesis. "In the beginning there was the astronaut," and the rest is created in his or her service. Skip the land and sea, trees or animals: MELiSSA creates a universe of algae, bacteria, plants, and humans.

I ask Gòdia Casablancas if this loop feels oppressive,

and he nods. "It is a completely domesticated ecosystem. But there's no other choice. People associate a deep-space mission with freedom and adventure, but it all takes place in a cramped, controlled environment, with very limited freedom."

I LOOK BACK AT THE CHART ON THE WALL, at the circle with arrows, colors, lines. The cycle of life, death, and regeneration reduced to a pocket-sized diagram. There is a paradox at the heart of this project. Yes, MELiSSA reminds us of our deep connectedness to Earth, but it also minimizes that connectedness by reducing it to a handy travel version of our biosphere.

That miniaturization of the biosphere is just temporary, says the Catalan professor when I present him with this paradox. "Our goal is not to leave Earth for good. We're not out to colonize planets. We want to travel further than we ever have, but then return with new insights."

MELiSSA, he emphasizes, is not about escaping, but rather about the scientific benefits of such a journey. Not only during the mission itself, but also—perhaps most important—throughout its preparation.

As we walk past the various research labs, he lists off the projects here on earth that are the fruits of MELiSSA-based knowledge. Water purification in Belgium. Algae farming in Congo. Urine filtration in French hotels.

After Gòdia Casablancas has explained almost a dozen machines and devices, we head to the university's cafeteria for a cup of coffee.

"What added value is a space mission to the MELiSSA project?" I ask. "All this research into how to survive in deep space mainly translates into adaptations that help us survive on earth. Shouldn't the project's mission be reassessed?"

Gòdia Casablancas shakes his head. "I know what you mean, but humans are simply fascinated by space. The notion of exploring the unknown, flinging ourselves into the darkness—there's nothing more exciting. I think one reason our research has reaped so many rewards is that it's carried out in the context of a future dream. We need that dream. Not only me, but the entire team. It's magical to work on something unimaginable. Even when you have no idea if or when it will actually happen."

He takes a sip of coffee. "These days, it can be hard to rationalize our focus on space," he says. "Some people feel we could better concentrate on solving problems here on earth. I can see that, but I also believe that you perform better research, learn more, and dig deeper, if your goal is something intangible. Humans are inquisitive by nature. We are driven by the desire for knowledge. And the universe is perhaps the greatest mystery of all, so yes, that is our major motivation. A human presence on Mars is a grand, emotional story about who we are and what we're capable of. Sending a robot there is a technical triumph; sending a human there would be a triumph for the soul, the spirit, the brain. A human can reflect upon his journey on behalf of all of us."

In the podcast *The Habitat*, Roger Launius, NASA's chief historian, reminds us that we've been planning a mission to Mars since the very beginning of manned space travel. We are always going to Mars "thirty years from now."

There is something comforting about that thought. A

perpetual dream is perhaps better than one that is fulfilled. You try to execute a "mission impossible" and in the meantime you discover the incredible benefits of spirulina and your connectedness with beets; you learn to purify urine and recycle water.

BACK IN THE HOTEL, I FaceTime with the home front. "Look, Mama is on Mars," David says. The children don't believe a word of it, they've seen *Secrets of the Red Planet*, an ESA animation film in which spaceman Paxi goes in search of alien life. "There's no air on Mars," the oldest says dryly. "And no bed, either," says the youngest, casting a critical glance at the hotel room behind me.

And they're off, *Dinotrux* beckons. We wave, blow kisses, and hang up.

I buy a takeout falafel from next door to the hotel and sit down with a book I brought with me from home: *The Case for Space* by the aerospace engineer Robert Zubrin, a well-known American advocate of Mars exploration who is convinced that humankind's only chance of long-term survival lies in leaving Earth, at least partially.

His point of view is at odds with that of the MELiSSA project. Just going to Mars is not enough, he says—we have to stay there. And for an undertaking like this, it's not humans that must be adapted, but the planet itself, through terraforming.

His book describes what needs to be done to make Mars habitable. The temperature—now an average of negative sixty-three degrees Celsius—must be increased, the atmosphere made denser, a protective ozone layer created. He suggests pumping greenhouse gases into Mars's thin atmosphere, which will raise the surface temperature and in turn vaporize some of the ice cap on the south pole, introducing extra carbon dioxide into the atmosphere and further jacking up the temperature. In half a century, he believes, the temperature on Mars can be raised by some fifty degrees Celsius.

Other scientists say it will take more like eight centuries; yet others think terraforming on Mars is insane and unfeasible. But they do agree on one thing: in theory, at least, it's possible. It would be like going back in time, because Mars was once (about four billion years ago) probably a warm, wet planet. I watch an animated film in which prehistoric Mars, in vibrant green and blue, transforms into the inhospitable, rust-colored planet we know today.

It is difficult not to see an analogy with Earth, not to think that future generations will talk about our planet in terms of "once." The Earth was once human friendly, with clean water and air. The climate was once hospitable.

I imagine a tragic relay race: humans terraforming one planet after the other. Destroying one ecosystem with greenhouse gases, only to use those same gases to resuscitate another, long-dead, world. And yet, there is something contagious about Zubrin's intergalactic optimism. It lacks the humility of cosmological awareness, but it sheds light on a horizon that far surpasses the doomsday scenarios we project onto Earth. What he preaches runs counter to the sinking feeling that's been dragging me down ever since our last heat wave. Zubrin believes we're living not at the end of time but at its beginning. He sees us as the forebears of an interplanetary species, rather than of a race that is self-destructing. He believes in our grand escape, that we will build a new world. And if we don't, our children will. Or their children. Mars is waiting for us, he writes. An ancient, dead world will, thanks to us, be brought to life.

Darkness has fallen outside, I set the book aside and close the hotel-room curtains. While I'm brushing my teeth I think of the MELiSSA project, a preparation for deep time as well as deep space. To engage with Mars is to engage with our future. Mars teaches us to think about ourselves in the long term. So now my quest for cosmological awareness has taken on another dimension. Space offers us another perspective not only on Earth and on ourselves but also on time.

12

•

The Present Matters Less and Less

THE LETTER *J* GLISTENS IN THE WET PAVING STONE ON THE Oudegracht, the main canal running through Utrecht, the Netherlands' fourth-largest city, home to a medieval city center and the country's largest university. Next to the *J* is a slippery *E*. Together, they form the first word—*Je*, or "you"—of a poem carved in stone and stretching the length of the canal. The complete first sentence translates to: "You have to start somewhere in order to face the past; the present matters less and less."

A sentence that took eighty-seven stones, more than a year and a half, to complete. A new letter is added every Saturday. One letter per week, three years for a stanza. This is the tempo of the poem, which has been creeping down the sidewalk since 2012 and eventually, two centuries from now, having snaked its way through the city, will arrive at Utrecht's Central Museum. The idea is that new generations of poets will keep the project alive. Like an inextinguishable Olympic flame, a chain of rhythm and rhyme that connects the present and the future.

Reading the lines requires you to progress at a snail's pace. Too fast, and the language breaks apart. I'm reminded of the Jewish myth of the shattered basin of light. The letters brought together like shards, chiseled into a poem for people we will never know. People who in future centuries will wander through an entirely different Utrecht from the one we know now—assuming the city still exists then. Maybe the letters will accidentally end up on some abandoned field, or at the bottom of the ocean. Or maybe the poem will be so long by that time that no one will know where it began. That ribbon of stone, where the language evolves foot by foot from an old twenty-first-century dialect into the tongue of future generations.

The drizzle turns into rain. People hurry to take cover under awnings. I try to imagine someone in the year 2520, walking along this poem and trying to imagine *us*. People from a time when the world revolved around burning eons-old fossilized matter. People who incinerated the past, and with it, their future.

I CONTINUE ALONG THE LINE of poetry, determined not to miss a single letter. After the last letter (for now) is a paving stone with a brief message carved into it: CONTINUATION NEXT SATURDAY AT 1 PM. I have to laugh at this prosaic announcement. Farther along the word snake, future year markers—2200, 2300—have already been carved into paving stones, awaiting their portion of the poem.

I've walked along this canal many times, but for the first time I notice that the paving bricks are laid in alternating horizontal and vertical orientations. Many of the stones are chipped, some are cracked. Maybe that happened before it was an auto-free zone, or maybe even earlier, when horses trotted heavily along the canal. The wet paving stones glisten in various tints of brown, red, gray, and blue. The slower I walk, the more colors I can make out.

Sir Wilfred Patrick Thesiger, an Englishman who trekked through Oman in the 1940s, observed in his book *Arabian Sands* that the landscape was so monotonous you had to travel slowly in order to notice its beauty. The quicker you go, he felt, the duller it is. Only a patient eye can pick up the thousands of nuances. The scrubby vegetation. A snake darting off. The lines etched in the stones.

The reason I'm walking through Utrecht in the first place is because of another desert landscape: the one on Mars. To be exact, it's because of a treatise by Jacob Haqq-Misra, an American astrobiologist who hopes that the far-away red desert will offer humanity a second chance.

A few days ago, I read his vision for the future colonization of Mars. He cites the Utrecht letter project as an example of the kind of long-term thinking we humans need to practice. Now that technological progress brings us ever closer to "homesteading" on Mars, he writes, it's time to think about the moral issues of interplanetary resettlement.

He warns that we must take care not to bring the colonial mentality (read: get rich quick; short-term gains) with us; that we should see our presence in space in the context of the *very* long term. Centuries. Millennia.

Haqq-Misra already has come up with a term for this ethic: "deep altruism." Selflessly striving for the welfare of future generations, on a scale of a thousand years or more. Generations who are only remotely related to us, who will still carry some of our genes but will have forgotten our names.

In short, Haqq-Misra says, it's all about solidarity with the humankind of the future. If we do go to Mars, we must see it as an opportunity for a new beginning. A second chance for our blundering species.

And to do that, we have to learn to think on another time scale.

THE DAY AFTER MY VISIT to the Utrecht poetry project, I Skype with Jacob Haqq-Misra in Seattle, where he is a senior research investigator at the Blue Marble Institute of Science. His camera is turned off, so I have to make do with his profile picture: a thirtysomething with long black hair, leaning against a tree, smiling modestly at the camera.

Haqq-Misra seems surprised by my interest in deep altruism. Most of his published writing, he explains, deals with the physical aspects of space. In his field, and in his own work, ethics is secondary. But "space ethics" has interested him ever since he first began his studies.

Many countries and private aerospace organizations are keen on getting to Mars as soon as possible, but there is almost no discussion about how we're going to behave once we get there. And we're talking not about technical protocols but rather about the mindset of the first human settlers. "I believe this offers us an incredible chance to redefine ourselves. But in order to do that, we have to be aware of the moral aspects of this journey. It can't be about what's in it for us. We have to strive to create a new sort of person, and therefore make choices that will truly benefit this new species in the long term."

He means "new species" literally. Settlers on Mars will have to become completely independent of Earth. Not immediately, of course. Earth will need to supply the settlers with the resources necessary for survival, but the end goal must be self-sufficiency.

"It could take centuries," he says, "but in the context of human evolution, that's not so long. There will come a point when the Martians no longer identify with Earth. They might not even call themselves humans. Because how human are you when you're so far removed from Earth?"

I try to take in what he's saying, and hear him laugh at the other end of the line. "Does this sound really sci-fi to you? It's the most logical outcome! More logical than the idea that people on Mars will still consider themselves earthlings."

He's right. If permanent settlements are established on Mars, what country will they belong to? Which legal system will apply? The best, or at least the most interesting, option is to see them as entirely sovereign. Free to shape their new society as they see fit. Free from earthly errors and missteps.

But it will only work, Haqq-Misra thinks, if we go there with the proper mindset. "We have to really understand what it means to create a new world. To reinvent ourselves in the context of future millennia." I'm reminded of the Iroquois in Paul Bogard's book, who advocate weighing up the interests of future generations in everything they do. An unimaginable task in our current moment.

Jacob Haqq-Misra believes we can develop these skills.

In his writing he draws attention to the Letters of Utrecht as an example of how we can train our long-term thinking. And in fact the poem is just a sketch, considering that it's projected to last only centuries and not millennia.

Another project he describes does last millennia. The Clock of the Long Now, a monumental-scale mechanical clock, is designed to keep accurate time, uninterrupted, for the next ten millennia. It is an initiative of the Long Now Foundation, an organization established to foster long-term thinking, as an answer to the ultrashort cycles of quarterly reports, deadlines, trimesters, and government terms that now dominate our way of life. After my call with Haqq-Misra I've still got an hour and a half before I have to pick up the children, so I surf to the Long Now Foundation's website. The first thing I notice on the home page is the way they notate years: the foundation was established in "01996." That extra zero appears consistently throughout. A subtle addition that creates an enormous shift in my awareness of time.

The organization explains that that zero was added in anticipation of the Y2K "millennium bug," the dreaded global computer crash (which in the end never happened). But I can't imagine they didn't foresee the effect it would have. The breathing space it creates. The zero as a reminder of what's to come. The eye of a telescope that can visualize a future beyond the horizon of living generations. Of course, no one can say whether that zero will ever become a one, or more than that. But just the possibility of it is a relief to me.

The clock itself isn't the goal, says one of the engineers in a video clip on the website, it's about the thought process that its construction stimulates. The foundation wants us to think about the next ten millennia. That's ten thousand years. A timespan equivalent to how long it took us to transform from hunter-gatherers to frazzled iPad addicts. This way, our civilization is given as much future as past.

AS I SCROLL THROUGH THE SITE, a simple illustration catches my eye. A semicircle with six concentric layers. These are "pace layers," I read in the accompanying text. Each layer represents a different pace of life. The outer layer is "fashion": ever-shifting trends, the latest craze. Underneath this is the pace layer of "commerce," followed by "infrastructure" and "governance." The two inner layers are, respectively, "culture" and "nature." These last two are the slowest paces, guided by memory, integration, and stability.

Every layer has its own characteristics. Fast learns, slow remembers. Fast gets all our attention but slow has the power.

It's important, I read, not to let the paces get jumbled up. Each layer has its own pace of change. For instance, if politics (governance, the fourth layer) were to outpace fashion (the hurried outer layer), then it would become too superficial. If nature, the slowest layer, were to speed up, then the ecosystem will be thrown off-kilter and in turn disrupt all the other pace layers.

I SEARCH FOR THE ACTUAL LOCATION of the Clock of the Long Now and read that the project has been adopted by Jeff Bezos. How ironic. The last person who should be appointing himself the guardian of long-term thinking is the billionaire known for stiffing his employees, creating mountains of waste, and earning a fortune from both—a man, in fact, who embodies exactly the kind of colonial mentality that Jacob Haqq-Misra warns us about.

I close the Long Now Foundation tab in my browser and look outside at the pink light above the square. Is there any point, I wonder, in building that clock while the world is burning up? Thinking about a new civilization on Mars while, here on Earth, the sixth extinction wave has already begun? Busying ourselves with life's pace layers while urgent action is needed to combat social inequality?

Bezos's takeover of the clock pushes my cynical button. I don't believe we should entrust long-term thinking to people like him. Short-term and long-term must coexist. Whoever chooses one over the other is choosing only half of life. Short-term thinking suffices on an individual level, but we are more than that. The invisible tendrils of our being extend further than our physical body and connect us with what lives and grows in other pace layers. Algae, beets, trees, rituals, ideas. You can't squeeze that enormous web into the four-year plan of a government cabinet or the twenty-four-hour news cycle.

The further you zoom out, the physicist and astronomer Janna Levin once said on a podcast, the more connections you see. Levin's research into black holes deals with time frames of a few billion years. Everything that happened less than five hundred million years ago, she jokingly calls "local politics." Extreme zooming out gives her a "warm, fuzzy feeling"; it changes your outlook on practically everything. Knowing that gold is the result of the collision of two dead stars makes our monetary system a galactic phenomenon. Realizing that we are all products of the Big Bang gives our identity an extra dimension.

By and large, we allow our worldview to be determined by the few preceding centuries. But compared to the billions of years during which we slowly evolved into what we are now, recent history has had a minuscule influence.

Extreme zooming out hasn't inured Levin to day-to-day issues, but it does put them in perspective. As Levin puts it, "There is no meaningful difference between an instant and eternity."

Somewhere on the endless internet I find an animation showing how the world might evolve in the next 250 million years. The moon slowly recedes, the sun cools off some, Africa and Europe slide together into one single landmass. North and South America let go of each other. Australia moves northward and attaches itself to Asia, Antarctica winds up somewhere near Madagascar.

Today's large-scale geopolitical conflicts will sort themselves out in a continental slow dance. Here we are building walls and fences, while Europe and Africa have been slowly falling into each other's arms for centuries now. And Australia is sidling up to Afghanistan, while the Afghans fleeing to Australia are detained on islands like hardened criminals. Those shifting tectonic plates remind us that alongside the reality of hard borders, there is another reality. The reality of slow but steady transformation.

DAVID TEXTS ME that he is on his way and will pick up some "dino hair," as our children call stir-fried noodles, from the Asian restaurant nearby. I look at my telephone. Yikes, the day-care center closes in ten minutes.

Out of breath, I push open the gate to the playground and remember what the woman from the family counseling office told us a few months ago in relation to our youngest's temper tantrums. Teach him to put things into perspective, was her advice. Strangely enough, it was the first time I really thought about that term, to *put things into perspective*—in Dutch, to "relativize." To understand your own position by looking at it relative to something else.

I used to hate it when grown-ups told me to put things in perspective, to look at them in relation to the bigger picture. As though the small thing that was infuriating me was less important than bigger things. But it's not about being *less* important, I know now, it's about being *more* interconnected. Putting something into perspective doesn't negate it, because seeing how it relates to something else is about noticing connections—the opposite of negation. More like restoration.

13

•

A Shadow World within Reach

ZOOMING OUT STARTS TO AFFECT MY DAILY LIFE. AS I
race home after an appointment, I am conscious of the
slow-growing trees looking down on me from either side of
the bike path. It doesn't make me cycle any more leisurely,
but something about their slowness sticks with me for the
rest of the day.

The children's curiosity helps. They can follow, with
endless patience, a slug trail in the front yard. When we go
out for a walk they stop every few steps to examine something: a toadstool, a tuft of moss, a duck's footprint in the
mud. My anxiety about the brokenness in which we live is
just as present as it ever was, but underneath, there's some
play in it. It's loosening its grip on me. It's still there—but
with something else alongside it. The universe, eternity, the
unyielding mystery.

Looking at Earth in relation to the other planets in our
solar system, I've come to appreciate how mild mannered it

is. From Earth's perspective, everything outside our atmosphere is extreme: the distances, the landscapes, the temperatures. But this planet fits us like a glove.

This thought makes me feel, in a new way, at home where I am, safely inside our atmosphere, the perfect place for our lavender plant to grow, for us to breathe, talk, greet one another, for the wood of the bench where Bob sits in the winter sun.

I TRY TO TRAIN MYSELF to observe the slowness of processes around me. Nightfall, the wear and tear of the paving stones, the growth of the lilac. One afternoon, I take the children to the Vliegenbos, a patch of woods adjacent to our neighborhood, to look for the hundred-year-old horse chestnut tree that, according to an app I downloaded, is one of the oldest in the neighborhood. The children's interest in trees is slim, so I lure them with the story that here, somewhere in the bushes, lives a Parasaurolophus. Winter is receding; among the tree trunks it is windless and warm. As we look for the chestnut tree, I wonder to myself why they call this place a *bos*, a "wood," rather than a park. It's not that big. Maybe it's the irregular layout. The haphazard paths and the density of the growth, the fallen trees that don't get taken away but are left to rot. It's wilder than a park. Here, kingfishers, buzzards, and long-eared owls breed; foxes and martens prowl about at night.

During our first years here, I largely avoided it. Not only because I didn't have children to take out to play. Its name, Vliegenbos, "Fly Woods," gave me the creeps. A fly-ridden wood doesn't sound like the most inviting place for a picnic.

Only when I read the information sign at the entrance did I realize what a dunce I'd been. Turns out it's not named after a fly population but after the politician Willem Vliegen, a Social Democrat responsible for creating an urban green space in this once-poor neighborhood. Not only so people would have a bit of greenery to stroll in, but to act as a buffer zone between our neighborhood and the chemical factory that still operates on the bank of the IJ River. A factory, incidentally, whose round-the-clock humming sounds like a permanent colony of flies.

Reality has elbowed out Willem Vliegen's idealism: mostly I come across well-heeled residents of the dike farther up and newcomers who, like us, have managed to find affordable housing in Amsterdam-Noord and are recognizable by their messy hairdos, trendy reusable water bottles, and expensive strollers.

Last year I responded to an appeal to join a neighborhood hyacinth-planting group. The call went out via the district's Facebook group, with its broad mix of users. I expected a more heterogeneous turnout—longtime residents and yuppie imports—but the hyacinth project attracted mostly newcomers. "Green and sustainable" is usually something that attracts the electric-cargo-bike crowd. Social issues like housing shortages and poverty, not so much. As though we newcomers prefer trees to neighbors.

Hey, there's the old horse chestnut! Muscular wooden arms stretch in all directions like huge roots in search of soil. From close up, the gray-green bark resembles an arid landscape. I examine the deep lines in the cortex, trace a river of wood with my fingertip. Farther up, the eldest shouts that he's found Parasaurolophus eggs. I amble along the winding path toward the sound of his voice. When I reach him, he excitedly points to an old pair of discarded white sneakers in the bushes.

Along the dirt path we take homeward, some way from the streetlamps that have started to flick on, the vegetation seems to be slowly sucking up the light. Colors darken, dissolve into countless tints of gray until it's like a huge black hand swipes over the trunks and branches, leaving only dark splotches behind. For a brief moment, I stand still among the trees, surprised by a darkness I haven't experienced in the city before.

It's not exactly a dark-sky preserve, but still, this close to our house there is apparently room for night.

The next evening, David and I take the children back so they can experience the darkness. It's a short walk, but their fear of the spooky paths makes us a bit uneasy, too, and we hardly dare to leave the bike path. But we do eventually make a foray into the darkness. The children's voices quiet to a whisper, our pace slows, and instead of making noise, we listen to the sounds around us. A motor scooter, the rain on the leaves, the hoarse call of a heron. For a moment we seem to be less present, we blend into our surroundings. We are shadows among the shadows, indistinguishable from a tree or a bush. Strangely enough, the woods feel much bigger than when we walked through them in daylight. "The fire makes a circle of light for everyone," wrote the German poet Rainer Maria Rilke, "but the darkness pulls in everything." This is exactly what I feel. We briefly drift through a world with no discernible beginning or ending. Our small urban wood is grand, mysterious, unfathomable.

As we shuffle along the path I'm struck by contradictory sensations: we could be anywhere, and yet at the same time I am fully present in the place I am now. A branch crackles somewhere. Startled, we hurry back to the light. Such amateurs. But still . . . There is a shadowy world a stone's throw away. I only have to dare to leave the bike path more often, and overcome my fear of crackling branches.

THIS DISCOVERY OF DARKNESS gives me a pleasant feeling of space in the days and nights that follow. It's as though there's a hatch within my head that leads away from light and haste; it opens onto an extra room with high ceilings and windows with views different from the one I've known until now.

This experience of space has an effect on the way I think. The more I zoom out, the more connections I see: between algae and astronauts, between myself and the crew of neighbors with whom I spin around the sun, day in, day out. I'm less irritated by the hubbub on the sidewalk in front of the teahouse at night, because I think I recognize the voice of the man who tends the lilac bush.

Maybe what I'm experiencing is an overview effect in slow motion. The distance that brings the world closer to me. And it's addictive. I want to go farther away, closer, out of our solar system.

14

•

Dwingeloo Galaxy

I'M IN AN EMPTY TRAIN CAR, WHOOSHING THROUGH THE
empty expanse of Flevoland, a province created in the
1950s from almost entirely reclaimed land. It is the end
of a warm, wet winter, and a feeble sun hovers above the
strange tundra-like landscape of the Oostvaardersplassen,
a twenty-two-square-mile nature reserve. In the distance
is a herd of grazing heck cattle, brought here to "rewild"
the region.

Rewilding—the word strikes me as self-contradictory.
Bright yellow intercity trains shoot through this "wilderness" and park rangers receive death threats if they don't
feed hungry grazers in the winter. The only wilderness we
still have is the one far above our heads.

The train exits the polder tundra, farther northward,
and brings me deeper into the cosmos than I've ever been:
out of our solar system. I'm on my way to ASTRON, the
Netherlands Institute for Radio Astronomy in Dwingeloo,

a village in the Dutch province of Drenthe. Here, every day, the facility collects radio signals from planets that circle a sun other than our own.

Exoplanets, they're called. Astronomers now have their eye on thousands of them. Planets with two suns, a planet made of diamond, rogue planets that, orphaned from their sun, tumble through space on their own. Incredibly hot gaseous planets. And planets that, like Earth, orbit in the "Goldilocks zone" of their sun: not too hot, not too cold. Just the right temperature to allow for liquid water and thus, possibly, life as we know it.

Perhaps, among these exoplanets, there is one similar to our own. One with roughly the same mass and a similar atmosphere. One that could be the Holy Grail in our search for exoplanets.

In his book *On the Edge of Infinity*, the physicist Stefan Klein calls it our potential "twin planet." I love this term. The romantic idea of an intergalactic twin, light-years away, that has created a different world with the same raw materials. A sibling showing us what we could have been, a mirror reflecting the varied possible expressions of our DNA.

But to find out, you have to look farther than our own planetary neighborhood. There's nothing in our solar system that reflects us back. The closest solar system that holds out any chance of finding a twin is Proxima Centauri, four and a quarter light-years away.

Another lovely term, "light-year." It sounds so tidy: not a heavy year, but a light one. A year you'd wish someone. In fact, it's a measure of time and space all in one, the distance light travels in a year: 5.88 trillion miles a year through frigid, mostly empty darkness. Now that I realize what a light-year is, I look approvingly at the blotches of sun hitting our wood floor. What a journey.

THE TRAIN CONTINUES through an old fenland under a wispy layer of cirrus clouds. My mother-in-law, who is babysitting today, texts me a picture of the children. I zoom in on their faces, the younger one's beaming smile and the narrow brown eyes of his older brother. It's strange how I always see them better in pictures than in person. In real life, they seem too close to really observe, and I pick up only fragments: a small hand, a cheek, two baby teeth. I love having my sons close to me, but only from a distance can I love them to the point of tears. Maybe that's what astronauts feel when they look back at Earth. Seeing something you love at a distance creates room for love that surpasses yourself, surpasses your own entanglement.

A lone taxi is waiting at the station. The driver swings open the door. "ASTRON?"

I nod. The only way to get to ASTRON from this station is by taxi. "Where are you from?" he asks in English. "Amsterdam." He laughs. He assumed I was from farther away. "The whole world stops here for a lift to the telescope."

He doesn't entirely get all that fuss about far-off planets, he says. He asks if I've ever watched the series *Ancient Aliens*. Before I can answer, he brings up the theory that the pyramids were built with extraterrestrial help. I nod occasionally, thinking about what I've read these past weeks about extrasolar planets.

The notion of other planets revolving around other suns

was for centuries no more than a theory. How could there not be, logically speaking—but evidence had always been out of reach. Until, in 1995, astronomers detected a planet orbiting, in a period of just four days, the star 51 Pegasi, fifty light-years from us.

From such a distance it's impossible to observe extrasolar planets directly, so they are studied indirectly through Doppler spectroscopy, or the "wobble method." Basically, the planet exerts a gravitational pull on its mother star, which then wobbles slightly. And the wobbling wavelengths of the light that star emits can be measured.

Other ways of detecting extrasolar planets include the "transit method," until now the most successful—and, in my opinion, the most poetic. When an exoplanet passes in front of its star (as seen from Earth), it causes a dip in that star's luminosity. This dip is barely observable, a decrease of around 1 percent, usually less. A cosmic wink.

In addition to using the eyes of optic telescopes, astronomers search for exoplanets with their ears as well. Telescopes that pick up radio or electromagnetic waves with frequencies different from the light waves we can see. And nowhere are scientists' ears peeled for exoplanets as keenly as in Dwingeloo, where ASTRON scientists analyze data collected by the world's largest radio telescope, LOFAR, short for Low Frequency Array.

I want to visit the place where the most incredible distances are brought within earshot, and meet someone whose daily task is listening to exoplanets: the ultimate act of putting things into perspective.

Of course, "listening" is figurative: the waves the scientists catch are not sound waves, which cannot travel in outer space anyway, as there is no atmosphere. LOFAR uses thousands of antennas to measure the radio waves of exoplanets, which a supercomputer then bundles together and turns into images.

I remember being slightly disappointed when I read that. Even though I knew better, I had hoped for something that humans could actually hear. The music of jingling stars, humming planets.

But to turn extraterrestrial radio waves into earthly sound requires human manipulation. I found the music for a duet for piano and Y Cam A, a pulsating star 1,100 light-years from us in the constellation Camelopardalis. The composer transposed the galactic frequency into an other-worldly lullaby.

You can listen to it online. The music is gentle and undulating. As though this ancient star, accompanied by the piano, has almost lost its voice. Around the same time, I discovered a project by the Dutch theater-maker and harpist Iris van der Ende. Several years ago, she wrote a piece for harp and stars. Melancholy strumming, while three stars, including the white dwarf GD 358, hum along. GD 358 is a dying star that in this composition produces a soothing, low-pitched sound.

Van der Ende's website also features a musical transcription of a "rapidly oscillating star," a star that expands and contracts every hour. The tempo is accelerated four thousand times, and still it sounds slow. A song from another pace layer, compatible with our pianos and harps only at high speed.

THE TAXI PULLS INTO THE PARKING LOT of a large building surrounded by woods and heathland. It's a pleasant, friendly-looking building, lots of wood in light tints. A school complex or an alternative care facility. I had pictured it as more exotic and alien. There's no sign of the radio telescope from here: its thousand interconnected antennas are spread throughout Europe. Every morning, these antennas feed the ASTRON researchers sixty terabytes of cosmic information. (To put that amount of data in perspective: sixty terabytes is enough to store fifteen million four-minute song tracks.)

At the reception desk, I'm met by astrophysicist Harish Vedantham, an athletic thirtysomething in hiking boots. He awkwardly sticks one elbow out to me. A new virus, Covid-19, is spreading like wildfire, and yesterday evening our prime minister went on national television, advising us to avoid close physical contact, followed by a goofy demonstration of this new pandemic handshake.

Following Covid protocols, Vedantham has placed two chairs six feet apart in his nearly empty office. Distance is my specialty, he jokes. Except that normally he doesn't think in feet but in light-years.

While he's off getting coffee, I look around his bare office. Nothing about it makes me think of the cosmos. The farther you get from Earth, it occurs to me, the less there is to see. I could at least look at the moon with my telescope; with Mars, there were those photographs taken by a remote robot. But with exoplanets, all we have to go on are artist's impressions. Illustrations showing what the planet *might* look like. Going into deep space requires imagination. And people like Harish Vedantham, whose research brings abstract distance nearer.

He returns with the coffee, takes a seat across from

me, and asks what I want to know. All sorts of questions shoot through my mind: I want to know if there's a limit to zooming out in the search for connectedness; whether my feeling that distance brings closeness is right; whether a twin planet affects our self-image; whether my search might in fact be a kind of escapism, as I sometimes fear. But I sigh and say: I would like to know what you are looking for out there.

"First, let me tell you about our routine," Vedantham says. He and his colleagues use the radio waves LOFAR collects and measures not only to look for new planets but also to study the magnetic fields of known exoplanets. "As far as we know," he explains, "a magnetic field is one of the basic conditions for life. You can't observe it with an optical telescope; for this, you need a radio telescope. It took us years to perfect the technique and properly analyze the data, but now things are moving fast."

No, they still have not found a twin planet. That's going to take a "lo-o-o-o-o-ong" time, Vedantham laughs. He's loath to make predictions, but okay, sixty years—forty or fifty if we're lucky. But it could also be a century before we know if anything is breathing and growing trillions of miles from us.

"The LOFAR data isn't enough," he says. "The presence of a magnetic field is only one piece of the puzzle. There are so many factors that determine whether a planet is habitable: its temperature, geological structure, makeup of the atmosphere, orbital velocity."

LOFAR's biggest success to date is the discovery of strange radio waves that Vedantham and his colleagues have identified as polar lights emanating from a red dwarf, a star smaller and colder than our sun. Polar lights, or auroras, are caused by the interaction between a star and its planet, so the light from the red dwarf led the researchers to a brand-new exoplanet that shares one thing with Earth: the aurora, or what we call the northern and southern lights.

Vedantham becomes animated when he talks about it. The breakthrough made global headlines, and he expects to spot another hundred or so of this kind of exoplanet in the coming years. A hundred new chances for cosmic kin.

The occasional success is important here in Dwingeloo, he says. "Progress takes time; that's just how it is when you're working on the outer limits of science. There are no well-trodden paths. I take comfort in the thought that our search is so significant, so revolutionary, that quick breakthroughs would almost cheapen it."

Revolution is the word astronomer Ignas Snellen used when I Skyped with him yesterday about the search for exoplanets. Snellen and his team at Leiden University study data collected by the Very Large Telescope in Chile's Atacama Desert, the world's largest optical telescope facility.

Locating a twin planet would be a "philosophical revolution" that would afford us a fundamental understanding of our origins. "Compare it to the theory of evolution," he says. "If Darwin had had only three animals—say, a parrot, a scorpion, and a cow—he never could have come to the conclusions he did. Think of Earth as an animal and you get the analogy: we need similar 'animals' to figure out our place in the big picture, to understand the chain our planet is part of." Since yesterday, I've had that image of Earth as an animal in my head. Part of a chain with plan-

ets we don't yet know of. A rare animal at that. Our solar
system seems to be unique: most seem emptier; not one has
been found with as many planets as ours. We could turn
out to be the life of the Milky Way party.

THE SUN COMES OUT, so Harish Vedantham and I decide
to continue our chat on foot. Walking across the heath,
we muse about the possibility of life other than our own.
The chance that a similar planet will be found is growing,
he says. Data from the now-retired Kepler Space Telescope
revealed a new exoplanet with the diameter of Earth, orbiting in the Goldilocks zone of its star. In 2018, NASA sent
Kepler's successor, TESS, into space to conduct a more
detailed search for exoplanets. And the European Space
Agency is also looking for our twin. In December 2019
it launched CHEOPS, a space telescope designed to study
the composition, mass, and density of four hundred to six
hundred small exoplanets to determine whether they are
gaseous or solid, earthlike planets. Then CHEOPS's successor, PLATO, will further investigate those falling into
the latter category. Each new project focuses on a different
facet, putting the pieces of the interstellar puzzle in place
one by one.

We pass the old radio telescope, the largest in the world
when it opened in 1956. Among its discoveries are two
small galaxies about ten million light-years from us: their
names, Dwingeloo I and Dwingeloo II, are a nod to this
dry bit of Dutch heath.

Vedantham asks what I'd prefer: to be alone in the universe or to know that other life exists somewhere. I don't
need to give it much thought. The idea that we are the only

beings in all that infinitude depresses me. Besides, it's too much responsibility. "And you?" I ask.

He thinks for a moment. "It would be awesome if it turned out there was life all over the place. But I'm also okay if we never find anyone and we're a completely unique phenomenon. The tricky part is that we have no idea of the possibilities. How do you define life? And how do you see what you don't know? We spend every day here looking without really knowing what we're looking *for*."

We keep walking. The heath spreads out around us in the hues of Jupiter: pale brown, yellow, soft orange.

I consider Vedantham's words. Aliens in books and movies always look a little bit like us. Ever since the invention of science fiction, aliens have been enlargements of ourselves, in better or worse versions. The Martians in *Two Planets*, a sci-fi hit from 1897 by Kurd Lasswitz, are peace-loving creatures who perceive us, through their gigantic eyes, as a cruel, small-eyed race. The aliens in H. G. Wells's *War of the Worlds* were exactly the opposite: malicious and stupid, with a far-too-large head and a feeble body; in fact, degenerate humans. And if the aliens were not misshapen humanoids, then they were a variation on some earthly animal. The heptapods in *Arrival*, a film based on a short story by Ted Chiang, look like giant octopuses. E.T. resembles a turtle, Alf an anteater, and the monsters from *Alien*, anglerfish.

How can you relate to something you can't imagine? If life on an exoplanet is so strange that we don't even know how to recognize it, if there is no relation to anything we know, then there's no way to put it into perspective. In the low sunlight of this winter afternoon, this thought makes me a little sad: that even if we do stumble across life somewhere else, there's a good chance we won't even notice it.

The sky begins to cloud over, the heather darkens. Vedantham and I walk back to the ASTRON building. In the parking lot we say goodbye with awkward little bows. As I wait for my taxi, I look at the clouds and wonder how far my gaze actually reaches. And if somewhere on the other side of the universe, some strange civilization has pointed a telescope at Earth to try to figure out if there's anyone here. What a bizarre idea: that we're a minuscule dot in someone else's sky.

I CAN'T GET THE ALIENS OUT OF MY HEAD. The day after my visit to Dwingeloo I delve into the Breakthrough Listen initiative, an international project in which hundreds of scientists are teaming up to search for alien civilizations on exoplanets.

I read a paper by the anthropologist Claire Webb, who specializes in the search for extraterrestrial intelligence. That is, she studies how the search is conducted. She watches scientists watch the cosmos. If you want to observe extraterrestrial life, she says, it helps to treat what you know as something alien.

In short, to recognize the unknown—in this case, alien life—you have to make it recognizable one way or another. Our brain simply refuses to see anything it doesn't expect, because it lacks a point of reference.

Think of that well-known film clip of a group of people throwing basketballs to one another. You show it to someone and then ask how many times the ball was thrown. They answer—eleven . . . no, thirteen . . . no, fifteen—and then you ask if they saw the gorilla. What gorilla? Well, the gorilla that walked straight through the group, which they didn't see because their attention was focused on counting the bouncing basketballs.

To be able to see what you don't expect, it's necessary to develop a kind of alertness, a way of looking that picks up irregularities. This is hard to practice, but you can turn it around, says Webb. You make what's normal strange, make the known unknown.

Look at the world the way an alien would. Scientists hoping to recognize alien life must first unlearn themselves and look anew. This can create a whole new concept of what a person is. What a "self" is.

Because, who knows, alien intelligence might have a completely different sense of time, or might see us not as individual beings but as part of a chain of growth and decay. Seen through their eyes, we might not be the autonomous creatures we think we are, but a temporary concoction of universal building blocks. Carbon, magnesium, iron. Universal elements that, for the duration of our life, are mashed up together with memories and stories into what we call human beings.

LATER THAT EVENING, I stroll around the neighborhood and try to look at familiar things through alien eyes. The front yards, the sidewalk, the rows of parked cars, the dogs and their owners out for a walk, the teahouse with the group of noisy regulars on the stoop, the bulky waste that gets dumped on the street every day even though it's picked up only three times a week.

Look with alien eyes. I repeat it like a mantra. But no matter how hard I try, everything still looks normal to me. Normal and—my heart leaps a bit when I realize this— much more familiar than it ever did.

15

•

Nowhere, Somewhere, Everywhere

THREE DAYS AFTER MY EVENING WALK, THE GOVERNMENT announces a nationwide lockdown. No more visits to radio telescopes, no city trips to Mars laboratories. The evenings are long and empty. Events have been moved online, but I can't face it. All those appearances and talks and discussions on Zoom.

I prefer to look up artist's impressions of exoplanets online, especially the ones where scientists believe life might be possible. The artist takes the information available and then fleshes it out with their own imagination. Half science, half imagination. A light purple Proxima Centauri, a brick-red Kepler-62f, a blue-gray TRAPPIST-1b (discovered by a team of Belgians). All those worlds we'll never visit.

Yet maybe they will still have something to remember us by: the "Golden Records," blueprints of humanity floating through interstellar space. Identical gold discs attached to *Voyager 1* and *Voyager 2*, NASA probes I spent my lockdown studying. In the summer of 1977 both *Voyagers* embarked on a scientific mission deep into our solar system. After they passed Jupiter and Saturn (*Voyager 2* even went as far as Uranus and Neptune), the giant planets' gravity flung the probes into interstellar space, where they now shoot through the cosmos at a speed of nearly ten miles per second, or thirty-five thousand miles per hour.

The Golden Records contain a message from us here on earth: our words and songs, our animals and landscapes etched onto an LP made to last a billion years. A "best of the blue planet" in ninety minutes of audio and 118 coded images.

The *Voyager* probes' cameras (which allowed NASA to produce the first-ever images of distant gaseous planets) have now been shut down, but much of the remaining equipment is still functional. At a distance of 14.5 billion miles (*Voyager 1*) and 12 billion miles (*Voyager 2*) from Earth, the probes are still sending us data.

The Golden Records' journey, one of humankind's most poetic missions, helps me through the disorienting first phase of lockdown. On NASA's website you can listen to the music that the small team of scientists and artists, led by the astronomer Carl Sagan, compiled as signs of life for alien listeners: Beethoven, Azerbaijani bagpipes, a Georgian protest song, Blind Willie Johnson's moving rendition of "Dark Was the Night, Cold Was the Ground."

It is an outlandish project, of course, incomplete and megalomaniacal. How can you represent all of life on earth in ninety minutes of music and 118 pictures? Sagan's team decided that it was important to make a good first impression. So there is no famine, war, or violence; no frowning; no chickens in factory farms. Blind Willie Johnson's voice, but not the segregation that caused him to die in utter poverty.

Yet I still find comfort in listening to these tracks that are now floating far outside our solar system. The Senegalese drums, the Peruvian panpipes, the greetings to the universe in fifty-five languages. Maybe because it all exudes so much optimism. What would we include today, more than four decades later, if we were to launch a new Golden Record? It would help if we had a recipient in our sights. Someone or something to send greetings to.

I surf to the Very Large Telescope's webcam. In a drawn-out panorama shot, the camera's eye glides across the empty observatory premises. The only reminder of human presence is a carelessly parked golf cart between the white buildings, which in that desolate terrain make you think of an alien settlement.

The night sky above the Very Large Telescope is the clearest in the world. Light sparkles everywhere, but the huge mirrors that normally focus on the universe are now sealed inside the domes. Because of the pandemic, there are no personnel to operate the telescopes. Our eyes to the universe are closed.

IF COVID-19 HADN'T MADE THE LEAP TO HUMANS, right now I would have been packing my bags for a trip to Piła, a small city in remote northwest Poland. A group of selected scientists, engineers, and medical experts were to gather in a dome, shut off from the outside world, to participate in a study of human behavior in isolation.

The dome belongs to LunAres, an analog research station for simulating crewed space missions as part of complex research into the psychological and physiological impacts of spending long periods in space. The plan was to spend two weeks as a participant, taking a couple of hours per day to work from my "moon office," as the project leader described it. I wanted to *really* try to imagine myself removed from Earth, to abandon myself to the fiction of a long-distance mission.

But Poland has gone into lockdown; internal European borders have been shut. The isolation I was so looking forward to has been replaced by another, sloppier isolation. Schools and day care are closed because of the pandemic, grandparents aren't allowed to babysit, David hurries to an empty newsroom every morning to meet a deadline.

I take a stab at homeschooling. I chuck the eldest's kindergarten curriculum overboard. We'll start at the beginning, I decide. The Big Bang. We listen to its echo on YouTube—cosmic background radiation converted into sound. An irritating, high-pitched hum that becomes lower and slower, turning into a rumble. It vaguely reminds me of sleeping on the back seat of a moving car.

The boys ask a thousand questions about the universe, most of which I am unable to answer. Where did the Big Bang come from? Why do we revolve around this sun and not another star? Why is space dark and stars light?

We look at the Hubble Ultra Deep Field—the photograph of the gleaming shards where my whole journey began. "Totally broken," the youngest concludes. I nod. They want to know how the solar system works, we try our hand at making a model, but the strings keep getting tangled. The paper-wad planets lie scattered on the table like a cynical message for the universe.

On the evening news, the Dutch astronaut André Kuipers, who has experienced long-term isolation firsthand, offers quarantine tips. Be open about anxiety, listen to one another, keep sleeping and moving. Suddenly, everyone feels like an astronaut on Earth, but not in the way Buckminster Fuller meant. Not like David Foster Wallace's fish in limitless water, but rather, like sardines in a can.

SPRING EVENTUALLY ARRIVES. News alternates between the pandemic and the massive wildfires in the US and elsewhere. A United Nations report warns that the climate crisis is unfolding in a series of worst-case scenarios. A local protest group plasters the city with flyers reading, "We can't go back to normal, because normal was the problem." Anti-vaxxers put "Fake News" stickers on top of these. An angry neighbor covers all the garbage bins and lampposts in the neighborhood with homemade stickers. White background, black block letters: IT'S ALL WRONG.

"What's wrong?" I ask him one day, as I pass him while he's at it. "Everything," he says. "Everything's wrong." His answer swirls around in my head as profound truth and utter nonsense all in one.

The world seems to be holding its breath. I never knew the square could be so quiet—aside from the birds, nothing around our house moves. The only car that passes with any regularity is the Red Cross van that picks Bob up every morning for his daily visits to Amsterdammers who have fallen on hard times during the lockdown. They distribute free groceries and provide transportation for the elderly and vulnerable. Whenever I see him leave his house in his white Red Cross coveralls, gloves, and a mask that covers half his face, it feels like fantasy has overtaken reality.

Bob, astronaut on earth. Ready for his mission on a dangerous planet. He looks lonely in his PPE. But everyone looks lonely these days. We're spacefarers in reverse. Up there, distance evokes feelings of closeness. Down here, closeness brings anxiety, and thus distance.

In the supermarket, the omnipresent circle of social distancing confuses me. Just as shifting celestial bodies can make a planet veer from its orbit, I lurch down the aisles, mumbling apologies if I accidentally breach the six-foot radius. It's got me muddled: I feel cooped up and at the same time yearn for confines. A grid where I can set the coordinates.

IT OCCURS TO ME that maybe I'm suffering from an earthly form of astrophobia. I've come across this term before, but only now do I understand it. Fear of the insanely big nothing we're whirling through. I realize that the terms I use to describe the spatial universe are pure nonsense. There is no "far corner" of the universe. There is nothing "surrounding" us. In infinity, there is no here or there from which you can determine your position. You float without coordinates in a cosmic abyss. Terrifying.

I read stories online about people who suffer from astrophobia. On a forum last year, a streetcar driver, Pete, wrote that every time he braked and came to a stop, he felt the universe fling him around some massive, freezing blackness. Paralyzed with fear, he ran his route through the city until arriving at the depot, trembling, at the end of the day.

I want to ask him if he got over that fear, and if so, how. But by now the forum has gone offline. And I imagine that these days, in the middle of this lockdown, Pete drives his empty streetcar through the city, with no one at the stops for some reassuring small talk. An anxiety-ridden streetcar driver in a mind-bending universe.

I walk to the window, look at the square, the café, the bench, the lilacs, and directly below, our front yard, where the spring colors are carefully coming out. The phlox, the viburnum, Bob's lavender.

You can counter the feeling of being nowhere, I think, only by being *somewhere*. By associating yourself with wherever it is you are. It suddenly feels ridiculous that I don't know where I live. My house, yes, but not the ground underneath it, not the depth of my garden. My nearest bit of earth.

I TAKE A SHOVEL from the balcony cupboard, go downstairs, and pry up some paving stones. I find sand, ants, beetles, cables, pipes, worms, all sorts of tiny animal holes. And then: water. I know the water table in this neighborhood is high, but still I'm taken aback. How can you live on top of so much water, so close by?

When I've reached the limits of where my shovel can take me, I read further. I read about the clay on which my family and I live, about the crushed shells surrounding the pilings that hold up the house. The many feet of sand that glaciers from Scandinavia pushed onto our street, which later formed tundra. Forty yards under our house is ground where bears, mammoths, even rhinos once walked. Woolly rhinos where the lavender now is.

Further down, deep under the tundra and the sand: rock. And yet deeper, the relatively thin earth's crust, then lava cores. And beyond them, past the center of Earth, a mirror image: again the crust and layers of ground, and above that the Pacific Ocean, exactly on the opposite side of the world, and thanks to the countless connections between here and there, part of the place where I am.

David laughs when I tell him later about my journey into Earth. "Going underground now for a way out?" I shake my head. I don't want to escape from where I am, I just want to expand it. That evening, for the first time since the coronavirus measures took effect, I don't feel so isolated. It's nice to know where I am. To have the feeling that I am somewhere, and via this somewhere, everywhere.

16

•

A New Yet Ancient World

SPRING FEELS LIKE SUMMER: HOT AND DRY. CONFINED TO the house and its cramped environs, I experience the season more intensely than ever. Everything that grows seems to exhale and burst open at the same time. Blossoms and buds everywhere. The cherry tree creates a light pink roof above the new sidewalk café, which is closed because of corona.

After being indoors for a few weeks, we install ourselves in the front yard. Together with Bob and the neighbors to our left, we've crammed our shared space with nectar plants to attract butterflies and bees. When an early heat wave sneaks up on us, we spend daytime indoors. Curtains closed, windows open. Another sprinkler ban, more of John's big, bare belly. Every morning, Bob hoists a red flag bearing the text WE'RE IN THIS TOGETHER in white letters. The fabric flaps in the summer breeze from morning till night. As if we are an autonomous republic: the United Citizens of the Front Yard.

I jump ahead in my homeschool curriculum, and to the children's delight we land at the era of the dinosaurs. This week's project is to make the meteorite that drove their favorite animals into extinction. It is a giant papier-mâché operation that so enthralls them that I finally have a few hours to myself. Enough to read, in one go, a book I ordered online: *Life Beyond: From Prison to Mars.*

In a project initiated by the astrophysicist Charles S. Cockell, detainees in Scottish prisons joined with astrophysicists to design a settlement on Mars. The book contains detailed logistics, technology, and infrastructure, but also a timeline wherein they think two centuries ahead, and on the Red Planet find solutions for Earth's most pressing problems. At the back of the book is a series of fictionalized emails written by various prisoners, who imagine themselves as crew members on board the first spaceship to Mars. "The reds are just so deep and rich, like spices; the shapes so familiar but alien for being untainted, fingers of waterless rivers stretch out across a new yet ancient world."

These are fragile stories of people feeling their way through uncharted territory. Not conquerors, but survivors. In their mind they shuffle across our cold, dusty neighbor. The only technology they have at their disposal is language and fantasy, and it's enough. As the poet Muriel Rukeyser said, "The universe is made of stories, not of atoms."

The book is just what I need to put my own feelings of confinement into perspective. If they can nurture their cosmological awareness in a Scottish prison, then I should be able to do it here, in my house.

I'M CURIOUS TO KNOW how the detainees look back on their space travel, so I email Cockell to ask if I could speak with one of them. His speedy answer is that it's not possible because of the strict lockdown but that he would be happy to talk to me. The boys' meteorite has now reached massive proportions, there is glue sloshing everywhere. "We're making one that can wipe out people, too!" the eldest shouts excitedly. I give him a thumbs-up and go upstairs to call Cockell in Edinburgh.

He tells me how thinking about Mars "repositioned" the prisoners in the universe. Being outcasts allowed them to more easily relate to a different reality than that of our everyday world. The unspoiled emptiness of Mars was precisely what made it so inspiring for them. An empty planet stimulates the collective imagination.

I ask him if something essential won't be lost when our physical presence takes over that emptiness. As long as space belongs to no one, it belongs to everyone. He thinks about it. "There's no hard answer to that. But we'll have to question our relationship with our solar system and make sure interplanetary travel doesn't just lead to more human hubris." He uses the term "cosmocentric ethics." An ethical system that includes the entire cosmos and all possible life forms, no matter how seemingly insignificant. The word *cosmocentric* seems to contradict itself. But maybe that's what is good about it: a word that reminds us there is no "center" except that one center that is all of us.

Downstairs, someone slips in a puddle of wallpaper glue; a slimy wad of papier-mâché splats on the upstairs landing. I quickly thank Cockell and hang up. The rest of the day is consumed with cleaning up the catastrophe of a gigantic domestic meteor impact.

THE NEXT DAY, I'm invited by a TV talk show to offer commentary on the launching of the *Crew Dragon*, a spacecraft built by Tesla tycoon Elon Musk's company SpaceX. It's a historic launch, the first time a private space agency will send humans into orbit.

The press has been focusing on it for days: this event marks, they write, the democratization of space travel, now that governments no longer hold a monopoly on it. For the media, the launch is clearly a welcome change from endless liveblogging about the pandemic.

But the jubilant tone bothers me. Elon Musk and his billionaire brethren Jeff Bezos and Richard Branson are modern-day space barons. Interplanetary aristocracy doesn't sound all that democratic to me.

That a couple of multibillionaires have broken the government monopoly does not mean that outer space is suddenly globally accessible to all. Thinking back on my talk with Charles Cockell, maybe even the opposite is true. If an empty planet symbolizes a collective brainspace, then colonization will surely deprive us of it.

Musk and Branson want to colonize Mars; Bezos, the space between Earth and Mars. "Control of space means control of the world," said Lyndon B. Johnson, the future US president, in 1958. Having invested billions of dollars in their private enterprises and made deals with national space organizations, the barons are well on their way.

Their language betrays a certain vision. Space as the last frontier, a place to colonize. As if that word doesn't buckle under the weight of its past. In an interview on this very topic, the American astronomer Lucianne Walkowicz argued against using the word *colonize* in an extraterrestrial context. Such a loaded word, they say, "both frames our future and, in some sense, edits the past."

Walkowicz, a specialist in the ethics of Mars missions, is afraid that space is becoming a place of exclusion. They prefer the term *inhabit*. And *manned* space travel, as far as they are concerned, can go straight into the garbage. Why not *crewed*, I think. New language for a new reality. I think of the Scottish prisoners, their meticulous sketches, their fragile stories that are so antithetical to the macho talk of those billionaires.

Still waiting for my answer to her invitation, the talk show's production editor assures me that their setup is "totally corona-proof." She apparently reads my hesitant silence as a fear of getting infected in a busy TV studio.

I tell her that while I'm interested in space, it's more from a philosophical, sociological, poetical perspective. A litany of *-icals* that usually makes producers hang up as fast as they can. But this is just what the editor seems to want. "Fine," she says, "the other two guests are pretty enthusiastic, so we could use a counterpoint."

The two other guests are men. Two male cheerleaders and one female killjoy—a recipe for regret. But I say yes anyway. I can't resist the thought of finally sitting in a room with strangers after weeks in self-isolation.

THE TALK SHOW BEGINS with a summary of the week's main news. Most of the attention is on the global Black Lives Matter demonstrations. SYSTEMIC RACISM IS THE

REAL PANDEMIC and DECOLONIZE THE SYSTEM are just two of the slogans regularly seen on placards. "Ironic to be talking about space travel now that everyone's forced to sit indoors," someone remarks between segments. I nod. But I see only the other, crueler irony: the dream of colonizing space alongside the dream of a decolonized Earth.

After three main segments, it's finally our turn. The discussion is short and is clearly meant as a lighthearted way to end the program. The overview effect comes up in the context of promoting space tourism, which the billionaires are keen to do. As if a few minutes of weightlessness and a quick glance at Earth will suddenly create global do-gooders. I'm given a minute and a half to summarize my doubts before the closing music kicks in. I can't find the right words. I get only as far as my qualms about the fact that one single earthman has a collective dream in his hands. The *Crew Dragon* launch, originally planned for that evening, is postponed due to weather conditions. I am—childishly, perhaps—secretly pleased.

I spend the whole next day going over and over in my head, as one does, what I should have said at that talk-show table. That it's possible to achieve an overview effect here on earth with a little bit more effort but a whole lot less rocket fuel; that all you need to do (for free) is to look up in order to be in awe of the place where you are; that there's no point in claiming unity in space if you ignore it on the ground; that, scientifically speaking, research into the overview effect is very thin indeed; that it's a wonderful story that can inspire us to look better, to think about distance, closeness, and connectedness, but there is zero guarantee that some random, rich space tourist will experience a spiritual epiphany from looking down at Earth for a minute; that it's perverse to use astronauts' experiences as PR sound bites for a multimillion-dollar industry and perverse to promote such a highly polluting industry as something that's good for our planet.

In search of more ammunition for my wet-blanket monologue, I come upon an interview with the astronaut Mae Jemison, a crew member on the space shuttle *Endeavor* in 1992. She calls the overview effect "a Rorschach test for what you believe in." Whatever you experience in space, she says, has a lot to do with how you see the universe— and your own existence—from here on Earth. When you look at it that way, it's not a view but a mirror.

TWO DAYS LATER, the rocket takes off anyway. Images of an emotional Elon Musk go viral. His arms raised in triumph, his eyes closed. A quote from the Danish theologian Thor Bjørnvig flashes through my mind: "Rocket fuel is a mix of kerosene and mythmaking." The euphoric Musk is a man of myths. If you weren't aware of the context, you might think that it was religious rapture, the arms stretched heavenward, the ecstatic smile on his face. Manned space travel as the ultimate achievement, side by side with the gods.

But if I've learned anything from my earthbound space journey, it is the simple fact that space is not "above" us. The universe surrounds us, is within us, it is the sun on our face, the moonlight and the aurora, the alien raw materials that make life possible and allow Bob's lavender plant to grow. And that is what makes space so awe inspiring.

I think back on my conversation with Jacob Haqq-Misra, the astrophysicist who advocates space ethics and deep altruism. Moving into space could mean a positive transformational experience for our species. But to have a transformational experience, you need to encounter something new. A new view, a new sound, a new landscape. If we fill in that landscape with our own historical baggage even before we see it, then it will never be able to surprise us.

In her book *A Field Guide to Getting Lost*, Rebecca Solnit reminds us that European colonists named American landscapes after the places they had left behind, forcing a place that could have taught them something new into an old mold. Instead of an exciting, new terrain, it was only a limp, disappointing copy.

In the end, it comes down to this: if you want to be changed by space, don't try to change it first. And in fact, you can substitute the word *space* in this context with just about anything. A view, a landscape, your neighbor.

17

•

In Light-Years There's No Hurry

IT'S A WARM SUMMER EVENING. I'VE JUST PUT THE CHIL-
dren to bed and go sit on the bench out front with a cola.
The yard smells of lavender after today's rain. It's funny
how long summer evenings tend to resemble one another,
as though conforming to the memory of earlier evenings,
the same scents, the same sensations.

I glance at my phone: Bob is late today. I've been waiting for him for nearly half an hour and am a little nervous.
I'm still not sure what I'm going to say to him. All I know
is that my space journey on earth won't be complete without sticking my neck out and trying to further close the
gap in our front yard. I have no idea if Bob is interested in
hearing my spiel, but I have to give it a try. I take a sip of
cola and look around me. Four large crows fight over some
old bread on the sidewalk; from the rooftops to the left, a
group of interested gulls looks on. A titmouse disappears
into the shrub, a sparrow shoots out of the summer lilac. I

follow all the flapping around me and think of the words of the pioneering Orville Wright, who said, "The desire to fly is an idea handed down to us by our ancestors who, in their grueling travels across trackless lands in prehistoric times, looked enviously on the birds soaring freely through space."

Without all these winged models, we might not have—probably would not have—taken flight ourselves. Just look at the *Eagle* that landed the first humans on the moon, SpaceX's *Falcon* rockets: beaks and wings have dominated the emblems of space missions for decades. My favorite mission patch is the white dove worn by the Soviet cosmonaut Valentina Tereshkova, when in 1963 she became the first and youngest woman (a record she still holds) to break through Earth's atmosphere.

Although the square is busy this evening—frolicking children, benches full of parents and grandparents—the relaxation of Covid regulations last week has failed to soften the national mood. Our divisions are still tangible, even here. Bob's solidarity flag still flaps from our house, urging the neighbors to be mindful of the precautions, while the grocer just fifty yards down the road still insists that the pandemic is one huge government conspiracy. We're divided into opposing camps—and I'm talking only about the virus. Alongside that, we've got irate farmers, angry at the nitrogen-emission measures, blocking highways with their tractors, and climate activists revving up for a week of late-summer demonstrations.

The controversies aren't solely terrestrial, either. This week I followed an extraterrestrial issue that has long divided humankind and has recently flared up again:

UFOs. The US Senate released a Pentagon report revealing that a portion of the American defense budget is earmarked for research into "unexplained phenomena" in the skies. A number of former employees told the *New York Times* they're sure that some of those came from outer space.

It's strange how many people simply do not want to accept that there are things that can't be explained, the religious scholar Diana Walsh Pasulka said in a *Times* podcast shortly after the report appeared. Pasulka is known for her book *American Cosmic*, in which she describes, among other things, the UAP ("Unidentified Aerial Phenomena," the Pentagon's new preferred term) sightings she found in medieval religious texts and contemporary ethnographic research.

She is surprised how skeptical scientists can be when it comes to studying UFOs. These kinds of sightings are so widespread, she says, that you would expect scientists to take them seriously. They don't. Even worse, it's nearly impossible to get financing for UFO research. See something inexplicable, and right away you're written off as a wacko. So people keep their sightings to themselves.

Likewise, here in the Netherlands there is deep-seated resistance to UAP research. I understand the skepticism: the UFO world is rife with conspiracy theories, and bizarre tales of alien abduction don't exactly add credence to the possibilities. But maybe the debate focuses too much on what UFOs might or might not be, and not enough on how we should look at them in the first place. If we would just accept that there's not an explanation for everything, then we might learn to better navigate the insecurity of our time.

The same thing feels true for so many topics these days.

It's as though every subject has only two sides, and they need to be fought tooth and nail. As though there's no such thing as "maybe" or "if" or "sometimes." Or a front yard with a lavender plant that offers completely disparate lives a common ground.

The only thing we know for sure, cosmological awareness tells us, is that everything is interconnected. Beet and spacefarer, moon and ocean, meteorite and humankind. Even things that, at first glance, look completely incompatible. Contradictions don't exist, they are only different coordinates in the same fabric. Cosmological awareness offers us an alternative to black-and-white: a different view from the thousand ways in which we divide land, sea, and people. Yes, we might have radical differences of opinion, but if we take a step back, we see how amazing it is that we even exist at all to have these disagreements. Accepting the improbability of our existence can make us take that step back, and maybe, who knows, cut each other some slack. "If a human disagrees with you, let him live. In a hundred billion galaxies, you will not find another," Carl Sagan said. Backing off for the sake of closeness.

BOB PULLS UP IN HIS GREEN MAZDA, waving theatrically, and parks right in front of our house. "It's going to rain soon," he says, "we'd better go inside." We sit drinking orange juice on his big leather sofa. He points to the aquarium, the tropical fish my sons love looking at. "My Sabbath rest," he says. "I don't go to church, but when I look at the fish, it's kind of like praying." We take a quiet moment to stare at the wrestling halfbeaks and the six-banded tiger barbs.

I ask Bob if he ever wanted to go into space. He nods. "I remember the first moon landing, all the excitement, the feeling that something that was once out of the question was suddenly possible. I wanted that, too."

He has never heard of the overview effect, but he can imagine the sensation. "I used to have dogs, I'd take them out for long walks, and I'd look at the sky and the clouds and feel completely at peace. Sometimes I took them to the beach, where we would walk until sunset, all those colors on the horizon, that's magical. Then you feel everything just glide off you. I'm guessing it's that feeling of vastness."

I ask if he can imagine being a spacefarer on Earth. "That's not so hard," he replies without hesitation. "Just look around whenever you're out: you see the sun, the moon, the sky, space. Walking and looking closely is enough."

I nod and think of everything I was planning to tell him about my space journey on Earth, about the loss of our night sky, about the moon tugging at us twice a day, about Rebecca Elson's "responsibility to awe" and my quest to come up with an answer to the vastness of space, about my botched trip to Lille and the algae that will keep us alive on our trip to Mars, how diffuse humankind's boundaries turn out to be, about all the pace layers we live in, how we have to make the extraordinary ordinary if we are to recognize alien life as such. But maybe it's better to just leave it at Bob's summary: Look closely. At the sky, at one another.

"And how do you share that feeling?" I ask.

Bob turns to me and says, "Now you're losing me."

Well, I begin, hesitantly, how could you use that experience to bridge the gap between yourself and someone who's not on your wavelength? I sigh. Why does it feel so vague, while what I mean is so simple? I try again. How could a sense that we're all earthbound space travelers bring people together? How do you share your experience? Not to erase all the differences, but to restore something and— I fall silent.

Bob turns back to the fish tank and sighs. "Well . . . usually my mind is a constant stream of: 'I've gotta do this and gotta do that, I, I, I' . . . But after a long walk, that's gone. I'm totally relaxed. And when you're relaxed, you're open. Then you can think. And listen. That's the point, right? Because when you're in a hurry, you're stressed, and then you don't see or hear anything. When you're not stressed, you've got the space to think about others, about their needs. And what you can mean in the bigger picture. So if I understand your question correctly, then my answer is 'Walk,' and if you want to share that feeling with someone, then you take them with you."

A ray of evening sun shines inside, I see countless tiny particles dance, and again I'm awed by the light that comes rushing at us at 670 million miles per hour. The LED lights in the aquarium change from yellow to purple to green. "New toy," says Bob. "Fun for the fish. You should've seen them this morning, darting all over the place to the lights. It was like a disco."

I think back on all the scientists, ideas, books, articles, films, and podcasts I followed this year. Looking at the fish in their watery nightclub, it's almost as though all roads inevitably led here. As though there was no other outcome than that via the moon, Mars, and Proxima Centauri b, I would end up in this living room, on this sofa. Via light-years to the here and now of this summer evening.

Last week I watched *John Was Trying to Contact Aliens*, a short documentary about John Shepherd, an American who has been obsessed with extraterrestrial life since he was a boy. He spent thirty years trying to make contact, broadcasting music into the cosmos from a radio transmitter in his bedroom. Never discouraged by the lack of a reply. On the contrary: Shepherd and his grandmother saved up for a larger "deep space transmitter."

The documentary is peppered with photos of John from throughout his life. You watch him grow older at the console as he searches, in vain, for contact.

And then, near the film's end, Shepherd is sitting on a sofa. He talks straight into the camera, tells how he met a man at a conference about extraterrestrial life. He looks to one side, the camera pulls back, he and the man sitting next to him look lovingly at each other. Then Shepherd turns back to the camera and says, "Contact has been made."

That I'm reminded of the film has nothing to do with romance. It's because of that last point, the whole point of the film: sometimes what you've been looking for light-years away is suddenly sitting next to you on the couch.

18

•

Night Watch

IT IS A RAINY AUTUMN NIGHT IN THE VLIEGENBOS. WET tree trunks glimmer in the darkness, the muddy floor of the woods sucks at my shoes. My neighbor Najat Kaddour walks beside me. Her father is the famous butcher whose shop recently moved from our square to larger premises down the street. Najat grew up in this neighborhood, left, and came back. "Noord's got such a great view of the sky," she said when I first met her.

"Walk," Bob answered when I asked him what the trick was to feeling like a spacefarer on Earth. "And if you want to share that feeling with someone, then you take them with you." That conversation with Bob got me thinking. The plan that gradually emerged is outlined on three sheets of letter-size paper pinned to the wall next to my desk. A Night Watch. A network of locals who take visitors to the Vliegenbos with a story about darkness.

An ecologist who grew up nearby will invite the "night

watchers" to see the woods through his eyes. He'll pass on the story of the fungi and the bats, and explain why darkness matters, and the night watchers in turn can share that information with new visitors. We'll build an observation platform in among the trees, where on clear nights we can set up the telescope. Quiet and darkness in an overlit urban neighborhood. And, most of all, space. Not only the actual space of a deserted patch of woods after dark, but also room for the neighborhood's trove of stories that can be shared at the post-walk campfire. Memories of darkness, stars, the night, and empty places now being crammed with posh apartment buildings and hotels.

SO HERE I AM, trudging through the mud with Najat and wondering how long it will take before we get folks from the neighborhood to join us in the soggy darkness. And I'm not even sure if it's "we"—I still have to ask Najat if she's willing to join my project. It's not only her energy and her interest in the stars and the moon that make her my ideal debut co-nightwatcher, but also that she's the daughter of the legendary local butcher and knows the neighborhood like the back of her hand.

I've organized this trial walk in collaboration with the nearby cultural center. Check which paths are suitable, how fast we walk without light, and where we can stop along the way. Our little test group includes Fred Haaijen, the ecologist who grew up in the neighborhood and played in the woods as a child. As we walk along the bike path, he tells us about the park's nighttime wildlife: the bats, the hedgehogs, the foxes. Then we go into the darkest part of the woods for a slippery walk to our final destination: a campfire that awaits us at the edge of the park.

It is the same dense, overgrown part of the woods that David and the boys and I hardly dared go into that first evening. There's a narrow opening between the two large elderberry bushes, a port of overhanging branches. We step through it one by one. The light from a few moments ago is now obscured by the trees and bushes. It's still far from being completely dark—the clouds still reflect the ambient light from the city—but still, it's harder to orient yourself than before. And there are no paths, the ground is littered with branches you have to step over so as not to trip.

Najat walks beside me. "Strange what this does to your senses," she whispers. She's right. The scent of rotting leaves seems so much stronger than during the day, the slightest sounds make you jump. Once my eyes are accustomed to the dark, I notice the countless nuances in color: for as far as I can see, the leaves on the ground are in all possible tints of gray. Leaf-shaped constellations on the black ground. A fine mist creates the strangest setting among the trees, accentuating their presence. Black figures looking silently down at us.

I'm disoriented, even though I know this section of the woods is small and we are near the edge. The darkness forces us to slow down. We slip-slide along the mucky ground; in front of me, someone stumbles over a root. There's cursing, giggling. Someone helps the person back up, and we silently continue our trek.

I look up. Between the treetops, in a moonless sky, I can count maybe twenty stars. I try to make out a constellation.

The Big Dipper, maybe? I draw imaginary lines among the points of light, try to invent my own constellation.

Looking up, I feel myself drawn into that age-old yearning to find answers up there to questions from down here. But constellations do not give you answers. On the contrary. As John Berger wrote, at night the stars "filch certitudes and sometimes return them as faith."

I think of my search for darkness in the Utrechtse Heuvelrug, and how relevant night seems to be in our present time, when so many of our certainties have fallen away.

NAJAT SQUEEZES MY ARM. "Wow," she whispers. I look around me and see the ground suddenly light up. A yellow glow that briefly breaks through the darkness and then disappears. It happens so fast that I wonder if I really saw it, but then there it is again. And again. All around us, the woods glow with the tiniest of lights. Glowworms. I had no idea they lived here, that so much natural light was hidden in these woods.

It's like these woods consist of dark matter, the glowworms are stars, and I am surveying the cosmos. The Dippers, Cancer, Sagittarius, our species' oldest stories glisten in the darkness.

I see myself suspended among the trees: a dot in the universe, a temporary form, descendant of a meteorite, countless organisms collected in a human skin. Bowled over by the vastness of the cosmos and by a handful of glowworms in a small urban wood in the Milky Way.

I slide my way further through the wet darkness. Our group has broken up, everyone following their own compass.

Najat and I proceed slowly. When we reach the bike path, it's clear we are lost: our route was supposed to take us *off* the path. The woods I know by heart during the day are suddenly terra incognita.

We randomly forge our way back into the growth, and I have a brief sensation that we are stepping back in time, and will spend years roaming through a prehistoric forest. But then I hear the murmur of human voices, and see in the distance, among the tree trunks, the flicker of a campfire.

Epilogue

A FEW SUMMERS HAVE PASSED SINCE THE ONE IN WHICH I found solace in the shards of the Hubble Ultra Deep Field. New exoplanets have been discovered, thousands of satellites launched, and a more advanced telescope has brought those shards into ever-sharper focus.

While our own atmosphere signifies a boundary, space appeals to our notion of boundlessness. The idea that, despite everything, we can still dream, grow, and exploit. This translates not only to awe, but also into the rapid escalation of space economy. Earth's orbits are becoming more and more clogged with space waste, and extraterrestrial mining plans are once again on the table. The issues of who space belongs to, and the consequences of putting a price tag on cosmic landscapes, are becoming ever more urgent.

In the reactions to my writings about space travel, I have noticed a growing divergence of opinion. There are those who see space ambitions as irresponsible as long as there's so much on earth that needs fixing. And on the other side are those who see space as an opportunity to escape a planet that, because of our own doing, has become increasingly inhospitable to humans. In between are the hedgers: the ones who hope we'll never have to choose between the close-by and the distant. That we can continue to dream about faraway planets while still caring for the one planet that makes dreaming possible in the first place.

The Night Watch has been growing steadily ever since Najat

and I walked through the darkness on that wet October night. Hundreds of people have followed us through our city forest after sunset. It has spurred a "night activism" that has even made its way to City Hall. There are plans to permanently switch off streetlamps in our district, and one of our night-watchers has been appointed "darkness advisor" to the city.

After more than two hundred night walks, I've learned that darkness has much more to offer than just an incredible view of the cosmos. I'm amazed at how much my world has grown since going on regular night walks. But that's for my next book. We still see very few stars, but we do talk about them, seated in small groups around a campfire, to combat intergenerational amnesia and keep reminding ourselves that a dirty pink night sky is not normal. We often close with a poem by Rafael Arozarena, who wrote that his legacy, expressed in earth, was a handful, but expressed in light was the entire universe.

Increasingly scant rainfall has made life hard for Amsterdam's glowworms. This summer (2022), Night Watch and the University of Amsterdam initiated an official glowworm count. For weeks on end, not a single one was sighted. But recently, after a heavy cloudburst, Najat and I saw a small group of larvae bravely glowing in the soil. They're still there. A few.

A Word of Thanks

I AM, FIRST AND FOREMOST, GRATEFUL TO THE ENTIRE PUB-lishing team of *De Correspondent*. Milou, Andreas and Channa, your faith and enthusiasm are the foundations of this book. Also, a standing ovation for Harminke Medendorp for her quick, smart, and sensitive feedback and her ability to stay focused whenever I flooded her with questions and doubts.

Anne Kuit, thank you for your enthusiasm in solving so many urgent questions. I have no idea what source you tap all that lively energy from, but it's such a privilege to work with you.

Thanks to Arjen van Veelen, Nina Polak, Lynn Berger, and Liet Lenshoek for their sharp eye. They were kind enough to read a first and very uncertain version of this story and comment on it. Merci, Roel van der Heijden, for rescuing me from some stupid mistakes, and Philippe Schoonejans, for your space-travel visions. Sahar Shirzad, your questions were much needed, and Christiaan Fruneaux, what would I do without a nerd friend like you?

This book would not exist without some important people who were willing to be part of my space journey on Earth: Samora Bergtop and Joos Ockels, thank you for the tea, the time, and your voices.

Bob Potemans and John Koopmans, I cannot say how much I appreciate your willingness to be part of your crazy neighbor's project. And yes, John, you know where to find me if you do not approve of this book. Najat Kaddour, our journey has been magical so far, and I am looking forward to all the light and darkness we will explore.

As always, thanks to David, my friend, my home; and to the two small space travelers without whom I could have finished this book so much earlier. I could not do without you for a day, Eyse and Otto. You keep me awake, and not just literally.

Thank you, Jessica Yao, for finding this story, sticking to it, and being so very precise. Thank you, Jonathan, for giving me a voice in a language that is not my own. And many thanks to Emma Parry for guiding me on this trip across the Atlantic.

Last but not least, thanks to the trees, from Sweden to Brazil, that formed the pulp that this book is made of.

Sources

EVERY STORY, THIS ONE INCLUDED, REVERBERATES WITH THE voices and ideas of others: philosophers, writers, friends, passersby. Some echoes emerge without your knowing it; others are consciously chosen.

The Dark Mountain Project, established by Dougald Hine and Paul Kingsnorth, was an important inspiration for this book. Explaining it here would require going into too much detail, but if you're not familiar with it, look it up, it's wonderful.

Another inspiration was The Nap Ministry, an organization, initiated by Tricia Hersey, that researches the liberating power of sleep. Rest as a form of resistance. This book does not contain concrete references to these projects, but their body of thought helped me to formulate my own ideas.

Then there is *A Field Guide to Getting Lost* by Rebecca Solnit (Penguin, 2006), which I reread several times while writing this book. I mention Solnit only once, and just briefly, but her *Field Guide* was my beacon, not only because of its themes, which sometimes touched on mine, but mostly because of the way she thinks through her ideas. The associations, the connections, the train of thought.

Additionally, there were, of course, concrete sources. When they are not explicitly acknowledged in the book, I've summarized them in the following, including the occasional extra lead for anyone wishing to explore in more detail.

All links were live as of September 2022.

1. LONGING FOR AN OVERVIEW

Dan McDougall's article on "ecological grief" appeared in the *Guardian* on August 12, 2019. For more information about "solastalgia," I recommend the 2005 article "Solastalgia: A New Concept in Human Health and Identity" by the Australian professor Glenn Albrecht, which was published in *PAN: Philosophy Activism Nature*.

The poem "Beyond the Bend in the Road" by Fernando Pessoa was translated by A. S. Kline (2018). Kline's translation can be accessed at poetryintranslation.com.

Images of the Hubble Ultra Deep Field, made by NASA's Hubble telescope, are plentifully available online. There is also a Hubble Deep Field, a montage from various exposures of the Big Dipper made by Hubble from December 18 to 28, 1995.

Additionally, there is the Hubble Extreme Deep Field, an image that zooms in on a specific portion of the Hubble Ultra Deep Field. All these images are well worth a look.

Tracy K. Smith's poem "My God, It's Full of Stars," about the Hubble telescope, comes from the wonderful collection *Life on Mars*, published in 2011 by Graywolf Press. The title of the poem refers to the book *2001: A Space Odyssey* by Arthur C. Clarke, first published in 1968 by the New American Library. These are the protagonist David Bowman's last words before he vanishes into a prolonged space journey. Smith's father worked on the construction of the Hubble telescope. Mike Wall's excellent 2012 interview with her can be found on Space.com ("'Life on Mars': Q&A with Pulitzer-Winning Poet Tracy K. Smith," May 4, 2012).

2. THE ATTITUDE OF AN ASTRONAUT

Brian Cox's quote, "We are the cosmos made conscious," comes from his BBC program *Wonders of the Universe*. The specific episode, from 2011, was entitled "Messengers." The complete quote is: "We are the cosmos made conscious and life is the means by which the universe understands itself."

A summary of what we know about phosphorus monoxide on the comet 67P/Churyumov-Gerasimenko can be found at eso.org, the website of the European Southern Observatory ("Phosphorus-Bearing Molecules Found in a Star-Forming Region and Comet 67P," January 15, 2020). An illuminating 2004 interview with Svetlana Gerasimenko provides additional details about 67P ("Happy Birthday: An Interview with Svetlana Gerasimenko," European Space Agency, February 23, 2004).

Buckminster Fuller's famous quote, "We are all astronauts on spaceship earth," comes from his book *Operating Manual for Spaceship Earth* (Touchstone, 1969). I encountered the work of Buckminster Fuller through Garry Davis, the man who issued world passports and started the world citizen movement after World War II. He was very much opposed to breaking up the world in bordered territories and used the astronaut's view of Earth as one system to support his movement. Fuller was a friend of his. The incredible story of Garry Davis has been mostly forgotten, but he was a true "astronaut on spaceship Earth." You can read more about him on the *New York Times*'s website (Margalit Fox, "Garry Davis, Man of No Nation Who Saw One World of No War, Dies at 91," July 29, 2013).

The first astronaut to cry on the moon was Alan Shepard, who was otherwise said to be a cold fish. It was the *Apollo 8* astronaut William Anders who said, "We came all this way to explore the Moon, and the most important thing is that we discovered the Earth" (the quote is paraphrased in chapter 2). In 2007, NASA published (on spinoff.nasa.gov) an excellent article about astronauts and their experiences, including Shepard and Anders, entitled "Detailed Globes Enhance Education and Recreation."

The video *Wubbo Ockels, Final Speech of an Astronaut* was recorded on May 17, 2014. It can be found on YouTube and on the Vimeo account of filmmaker Inge Teeuwen.

The *People* magazine interview "Edgar Mitchell's Strange Voyage," in which astronaut Edgar Mitchell (*Apollo 14*) made his famous comment about the politicians who should try looking at Earth from the moon, can be found online at people.com (April 8, 1974).

The interview in which Anousheh Ansari reflects on her trip to the International Space Station appeared in *Wired* on May 11, 2016, and was reported by James Temperton. Its title is "Anousheh Ansari Was the First Female Space Tourist. Now She's Inspiring the Next Generation."

To see the documentary about the overview effect, go to weareplanetary.com or go directly to YouTube or Vimeo. It is called *Overview* and was made by Planetary Collective in 2012.

The full title of Frank White's study is *The Overview Effect: Space Exploration and Human Evolution* (Houghton Mifflin, 1987).

The study "The Overview Effect: Awe and Self-Transcendent Experience in Space Flight" by David B. Yayden et al. (*Psychology of Consciousness: Theory, Research, and Practice* 3, no. 1, 2016) can be found at apa.org.

For David Foster Wallace's fantastic commencement address to the Kenyon College class of 2005, search YouTube for: "This Is Water—Full Version—David Foster Wallace Commencement Speech" (2013). The speech has also been published in book form by Little, Brown (2009).

3. EARTHGAZING AS THERAPY

I lowered my expectations of the Columbus Earth Center somewhat for the benefit of the narrative, but they were nevertheless far too high. For more pragmatic people than myself, I can wholeheartedly recommend a visit to the center in Kerkrade, the Netherlands.

Readers interested in learning more about Wallace J. Nichols's "blue mind" are advised to consult his book *Blue Mind* (Little, Brown, 2014).

Bruno Latour's 2017 theater piece in which he challenges the astronaut's perspective is called *Inside* and can be viewed online via Latour's website: bruno-latour.fr.

The podcast about the astronaut who wasn't that impressed by his trip to the moon was produced by the NPR program *This American Life*. The story, "The Not-So-Great Unknown," was reported by David Kestenbaum (episode 655, 2018).

Annahita Nezami's TEDx Talk, "The Therapeutic Value of the Overview Effect and Virtual Reality," can be found on YouTube—although by now, the talk (which she gave in 2017) is out of date. As of July 2022 Nezami was still developing the program. You can follow it on the VR Overview Effect website: www.vr-overview-effect.co.uk/.

4. SPACEFARERS WITHOUT STARS

On the subject of the loss of the night sky, I can highly recommend Paul Bogard's book *The End of Night: Searching for Natural Darkness in an Age of Artificial Light* (Little, Brown, 2013), which I quote in this and the following chapter. I also recommend Al Alvarez's *Night: Night Life, Night Language, Sleep and Dreams* (W. W. Norton, 1994), a quotation from which appears in chapter 5.

Another excellent book offering a historical perspective on our relationship with night is *At Day's Close: Night in Times Past* by E. Roger Ekirch (W. W. Norton, 2005). In it, he describes how before the invention of the electric light, people probably slept in two phases, separated by a brief period of wakefulness in which there was room for things that did not fit in during the day: storytelling, fire building, lovemaking. I like this idea. A long night, in which you alternate being asleep and awake in a kind of poetic twilight. But then again, I have a tendency to romanticize this kind of thing.

The article "The New World Atlas of Artificial Night Sky Brightness" (Fabio Falchi et al., *Science Advances* 2, no. 6, 2016) can be found online at Science Advances.

5. LIGHT AND NIGHT

More information about Kamiel Spoelstra's (and his colleagues') research in Wageningen can be found on the Netherlands Institute of Ecology's website: nioo.knaw.nl/en/projects/artificial-light.

The study of mice was conducted by Bert van der Horst, endowed professor of chronobiology and health, and research scientist Inês Chaves. The *Erasmus Alumni Magazine*, available online, features an interview in Dutch with Bert van der Horst called "Je biologische klok is van levensbelang" (reported by Marjolein Stormezand).

The study of the relationship between lighting and public safety was conducted by Barry Clark of the Astronomical Society of Victoria and is entitled *Outdoor Lighting and Crime, Part 1: Little or No Benefit* (2002).

Whoever hasn't yet read *The Hitchhiker's Guide to the Galaxy* by Douglas Adams (Pan Books, 1979) should do so at once. I can also highly recommend the film version directed by Garth Jennings (2005).

Thierry Cohen's photographs, in which a night sky is super-imposed onto modern cities, can be seen on Cohen's website, thierrycohen.com.

There is more information about Govert Derix and his book *Sterrenmoord* (Uitgeverij Tic, 2013) on his website govertderix .com.

Fernando Pessoa wrote the poem quoted at the end of this chapter, "From my village I see as much of the universe as can be seen," under the pseudonym Alberto Caeiro. The English translation is by Richard Zenith (*A Little Larger Than the Entire Universe: Selected Poems*, Penguin, 2006).

6. COSMOLOGICAL AWARENESS

Bringing Down the Moon was written by Jonathan Emmett (Candlewick Press, 2001). The Dutch version was published in 2011 by Van Goor. My children love this story.

If you would like to receive newsletters from ESA and NASA, you can sign up for them at esa.int and nasa.gov.

My favorite weatherwoman, the Space Weather Woman Tamitha Skov, has her own website: spaceweatherwoman.com.

Wil van den Bercken's terrific book *Uit sterrenstof gemaakt* (Made of stardust) was published in 2020 by Kok Boekencentrum. I interviewed van den Bercken for *De Correspondent*: "Antidote for thinking small: grasping the infinite universe" (March 18, 2020).

7. THE SECRET BREATHING OF EARTH

I advise all space enthusiasts to follow Maggie Aderin-Pocock. Not only does she carry out fascinating scientific work, but her lectures and television programs are inspirational and require little knowledge of the cosmos.

She is currently working on an animation series called *Interstellar Ella* for children from four to seven. The trailer can be seen on Apartment 11 Productions' website: apartment11.tv.

The "chin of gold" quote by Emily Dickinson (1830–1886) is taken from her poem "The Moon Was But a Chin of Gold." The phrase "white as a knuckle" comes from the poem "The Moon and the Yew Tree" by Sylvia Plath (1932–1963).

I encountered Wang Zhenyi's quote about how the moon makes no sound but can still be heard in Maggie Aderin-Pocock's *Book of the Moon* (Abrams Books, 2019).

Michael Collins's words following his solitary circuit above the moon are said to have been, "I knew I was alone in a way that no earthling has ever been before." Unfortunately it's not known precisely when or where he said this.

I recommend *La lune est un roman* (Once upon a moon) by Fatoumata Kebe (Slatkine & Cie, 2019) for anyone who likes the mix of science and poetry.

8. AN ANSWER TO THE DISTANCE

One of the greatest discoveries of my journey was the work of Rebecca Elson, on the fantastic website themarginalian.org, run by Maria Popova. Every year Popova organizes The Universe in Verse, an evening of poetry exclusively about science and the natural world. The website's archive contains recordings of previous performances, and one day I stumbled upon a reading of Elson's "Theories of Everything" by the singer Regina Spektor.

Elson's work is hard to find—this particular poem comes

from the collection *A Responsibility to Awe* (Carcanet Press, 2001).

The words of Ursula K. Le Guin that I quote in this chapter come from the foreword to her collection *Late in the Day: Poems 2010–2014* (PM Press, 2015).

Rebecca Elson's statement "Because curiosity, after all, is also of the spirit" also comes from her (sole) poetry collection, *A Responsibility to Awe*.

My information about the journey to the south pole of the moon was taken from NASA's website nasa.gov (Brian Dunbar, "Moon's South Pole in NASA's Landing Sites," April 2019).

More information about the permanent space station orbiting the moon can be found on the ESA website: esa.int. On the site, look up "Moon Village."

I refer those interested in mining on Mars to the website ispace-inc.com.

Jon Kabat-Zinn's quote is the title of his book *Wherever You Go, There You Are* (Hachette Books, 2005).

Information about the Outer Space Treaty can be found on unoosa.org, the website of the United Nations Office for Outer Space Affairs.

The quotation "Somewhere, something incredible is waiting to be known" is frequently attributed to Carl Sagan, but in fact was written by the journalist Sharon Begley for a profile of Sagan that ran in the August 15, 1977, edition of *Newsweek*. You can read more about the source of the quotation on the website Quote Investigator, at quoteinvestigator.com/2013/03/18/incredible/.

I would encourage everyone to become a member of the Moon Village Association: moonvillageassociation.org.

9. MUSEUM OF THE MOON

The installation *Museum of the Moon*, created by the British artist Luke Jerram, is at this writing still on a world tour (subject to Covid restrictions; see my-moon.org/tour-dates).

The multiple-choice question soliciting opinions on the value of developing long-term infrastructures on the moon is drawn from the Moon Village Association's 2019 survey about Moon exploration, conducted in partnership with the World Space Week Association. The language has been adapted for readability. You can view the results of the survey on the Moon Village Association's website: moonvillageassociation.org/are-you-ready-for-the-moon-village-survey-results/.

Items on the "canals" of Mars can be found in a variety of newspaper archives, but for a comprehensive discussion, I recommend the Dutch-language book *De kosmische komedie* (The cosmic comedy) by Frank Westerman (Querido Fosfor, 2021), which offers a wonderful introduction to the astronomer Giovanni Schiaparelli (1835–1910), the "discoverer" of these canals.

I can also recommend the book *Light in the Darkness: Black Holes, the Universe, and Us* by the German astrophysicist Heino Falcke (English edition published by HarperOne, 2021), in which he describes the slowness of light and the tangents between science and religion.

The interview with Heino Falcke (by Willem Schoonen) mentioned in this chapter appeared in the Dutch newspaper *Trouw* on October 31, 2020, entitled, "This is Heino Falcke, the astrophysicist who showed the world a black hole."

10. SUNSET ON MARS

Gimlet Media's podcast *The Habitat* (2018) is available via most podcast apps as well as the website gimletmedia.com.

You can follow the MELiSSA Project by registering for the newsletter at the website melissafoundation.org.

The study into the ideal number of astronauts on a Mars mission can be found at ieeexplore.ieee.org: Jean-Marc Salotti et al., "Crew Size Impact on the Design, Risks and Cost of a Human Mission to Mars," 2014 IEEE Aerospace Conference (June 19, 2014).

NASA astronaut Frank Culbertson, the only American who was not on Earth during the 9/11 terrorist attacks, tells more about his experiences in a video on Space.com: Megan Gannon, "Astronaut Frank Culbertson Reflects on Seeing 9/11 Attacks from Space" (September 11, 2017).

And for those who want to get addicted to Martian sunsets, mars.nasa.gov is a good place to start. Once on the site, go to "InSight Captures Sunrise and Sunset on Mars" (May 1, 2019).

11. BEAM ME UP, SPIRULINA

When Species Meet by Donna J. Haraway (University of Minnesota Press, 2007) offers some heavy-duty (and somewhat heavy-going) insights into our interspecies connectedness.

Roger Launius, the historian who said that we're always going to Mars "thirty years from now," made this statement in the bonus episode of the podcast *The Habitat* entitled "This Is Not My First Rodeo" (Gimlet Media, 2018).

The full title of Robert Zubrin's book is *The Case for Space: How the Revolution in Spaceflight Opens Up a Future of Limitless Possibility* (Prometheus Books, 2019).

One of the animations that show Mars drying up can be found on National Geographic's YouTube channel: *Mars 101* (May 31, 2018).

12. THE PRESENT MATTERS LESS AND LESS

Wilfred Thesiger's ruminations about traveling slowly so as not to get bored come from his book *Arabian Sands* (Longmans, 1959).

"The present matters less and less" is by the poet Ruben van Gogh. His complete contribution to the Letters of Utrecht (the first lines of the poem) is: "You have to start somewhere to give the past a place, the present matters less and less. The further you are, the better. Go on now," after which other poets continue the text. The Dutch original text reads, "Je zult ergens moeten beginnen om het verleden een plaats te geven, het heden doet er steeds minder toe. Hoe verder je bent, hoe beter. Ga maar door nu."

Jacob Haqq-Misra's articles can be found on his website: haqqmisra.net. In addition to future Martian ethics, he writes accessibly about many other topics, such as the search for extraterrestrial life and climate change on other planets. His book, *Sovereign Mars*, is forthcoming from the University Press of Kansas as I write this. I'm very curious to read it.

The Long Now Foundation website (longnow.org) provides a thorough description of the 10,000-Year Clock. Among the creators of the clock is Stewart Brand, who also formulated the idea of the pace layers. A critical note on this project that I thought was interesting can be found in the episode "The Real Legacy of Stewart Brand" on Paris Marx's podcast *Tech Won't Save Us*.

Janna Levin was interviewed by Tim Ferriss on his podcast, *The Tim Ferriss Show*, in an episode called "Janna Levin on Extra Dimensions, Time Travel, and How to Overcome Boots in the Face" (episode 445, July 8, 2020). It can be heard on (nearly) any podcast app.

The animation *How Earth Will Look in 250 Million Years* can be found on YouTube (Tech Insider, September 25, 2017).

All Dutch speakers interested in zooming out on time are advised to sign up for the Dutch-language newsletter put out by Studio Monnik, a small think tank founded by my good friend Christiaan Fruneaux and his associate Edwin Gardner. These "chrononauts" offer weekly new insights into the Long Now.

13. A SHADOW WORLD WITHIN REACH

A historical tree and greenery map of Amsterdam is available online at maps.amsterdam.nl/monumentaal_groen/.

You can view a map of light pollution around the world at www.lightpollutionmap.info.

A good overview of the initiative and activities about darkness and light pollution can be found on the website of the International Dark-Sky Association, darksky.org/light-pollution/. You can find a list of official Dark Sky parks, reserves, and sanctuaries on their site, too.

Rainer Maria Rilke's words are taken from the poem "You Darkness, That I Come From," translated into English by Robert Bly in *Selected Poems of Rainer Maria Rilke* (Harper & Row, 1981).

14. DWINGELOO GALAXY

Stefan Klein's *On the Edge of Infinity: Encounters with the Beauty of the Universe* (Cassell Illustrated, 2018) is an excellent primer on cosmological awareness. Well written, accessible, and chock-full of mind-boggling facts.

The program the taxi driver mentioned on the way to ASTRON, *Ancient Aliens*, is an American TV series made by Prometheus Entertainment (2010) in which human history is examined under the premise that there had once been contact between humans and an alien civilization. Excerpts from the series can be found on YouTube.

The duet for piano and Y Cam A was written by Burak Ulaş and can be heard via the *New Scientist* website: the article, by Joshua Sokol, is entitled "Listen to the World's First Duet for Piano and Pulsating Star" (August 12, 2015).

Iris van der Ende's *Stellar Sound Show* is on her website: irisvanderende.nl.

The science journalist Govert Schilling contributed a good item on the Very Large Telescope to the Dutch edition of *New Scientist*: "Welcome to star paradise Atacama" (November 16, 2016).

In the interest of brevity, this chapter does not mention Ariel, the ESA program that follows CHEOPS. More information on Ariel can be found at: sci.esa.int/web/ariel.

I recommend looking up some artist's impressions of exoplanets. Like the Martian sunsets, these images have an alienating quality to them. In an article for the *Daily Mail*, Cheyenne MacDonald describes how these impressions are created: "The Art of Alien Worlds: NASA Reveals How It Creates Its Incredible Exoplanet Artist's Impressions" (June 8, 2017).

This chapter briefly mentions the film *Arrival* (directed by Denis Villeneuve), where alien heptapods come to Earth with a message for humankind. It is one of the things I would have liked to write more about, but there was no way to incorporate it neatly into the rest of the story. Not to worry, the heartwarming heptapods will surely find their way into my work another time. I am envious of anyone who has not yet seen the film. *Arrival* came out in 2016 and is based on a short story by Ted Chiang, "Story of Your Life" (1998), which you can read in the collection *Stories of Your Life and Others* (Tor Books, 2002). The film score was composed by Max Richter, and I continually played his piece *On the Nature of Daylight* while writing this book.

The paper by Claire Webb is the dissertation she wrote for her doctorate at the Massachusetts Institute of Technology, titled "Technologies of Perception: Searches for Life and Intelligence beyond Earth." You can access it at dspace.mit.edu/handle/1721.1/129021.

The film of the group of people tossing a ball back and forth while a gorilla walks by is always a fun way to open a discussion into everything we don't see while we think we see so much. It can be viewed at theinvisiblegorilla.com, where you can read more about the experiment and the researchers, Christopher Chabris and Daniel Simons.

15. NOWHERE, SOMEWHERE, EVERYWHERE

Information about the dome in Poland can be found on the website lunares.space.

The podcast I made with Milo Grootjen, resident astronomer at Amsterdam's Artis Planetarium, can be heard (in Dutch only) via *De Correspondent*: "De thuisastronaut. Wat je ziet als je vanavond scherpstelt op de maan" (April 7, 2020). I recommend listening while looking at the moon. Milo also gives terrific courses on the universe. For more information, go to the Artis website: artis.nl.

At the website voyager.jpl.nasa.gov you can find every track on the Golden Records. Kurt Waldheim's moving speech is on YouTube: "The Voyager Interstellar Record—1/31 Greetings from the Secretary General of the UN Kurt Waldheim" (published on October 1, 2011).

Waldheim's quote "To be taught if fortunate" became the title of a book by Becky Chambers (Hodder & Stoughton, 2019) in which she offers a radical new proposal for future space travel, supported and paid for by all of humanity, driven by research rather than colonization.

The Muriel Rukeyser quote is taken from the poem "The Speed of Darkness," from *The Collected Poems of Muriel Rukeyser* (McGraw-Hill, 1978). You can read the entire poem on the Poetry Foundation's website: www.poetryfoundation .org.

16. A NEW YET ANCIENT WORLD

The interview with Lucianne Walkowicz can be found on the National Geographic website: "We Need to Change the Way We Talk about Space Exploration," by Nadia Drake (November 9, 2018).

Lyndon Baines Johnson, US president from 1963 to 1969, was a driving force behind the space race between America and the Soviet Union. A race he was later forced to put on hold due to the enormous financial cost of the war in Vietnam. More about Johnson and his influence on the aerospace industry can be found in "LBJ's Space Race: What We Didn't Know Then (Part 1)" by Alan Wasser for *The Space Review* (June 20, 2005).

Imagining Outer Space: European Astroculture in the Twentieth Century by Alexander C. T. Geppert (Palgrave Macmillan, 2018) is one of those rare books that looks at space travel not from a technical or scientific angle but rather through a cultural and anthropological lens. Another unique aspect is that it focuses exclusively on European aerospace, which has traditionally played second fiddle to the American space program.

It was my intention to visit the renowned astrobiologist Charles Cockell in the spring of 2020 in Scotland, but Covid-19 prevented that. As a fan of his work, I would have very much liked to meet him in person. His finest project is, in my opinion, From Prison to Mars, in which Cockell invited detainees in Scottish prisons to think about settlements on Mars. The results were brought together in a moving collection entitled *Life Beyond: From Prison to Mars* (British Interplanetary Society, 2018). Imprisoned on Earth, the convicts dreamed about our future on another planet.

17. IN LIGHT-YEARS THERE'S NO HURRY

The Wright quote could have come from either brother, but it is usually attributed to Wilbur. The complete quote is: "The desire to fly is an idea handed down to us by our ancestors who, in their grueling travels across trackless lands in prehistoric times, looked enviously on the birds soaring freely through space, at full speed, above all obstacles, on the infinite highway of the air."

The article about the Pentagon's investigation of UFO sightings can be found on the *New York Times*'s website: "No Longer in Shadows, Pentagon's U.F.O. Unit Will Make Some Findings Public," by Ralph Blumenthal and Leslie Kean (July 23, 2020).

Diana Walsh Pasulka was interviewed about UFOs by Ezra Klein for his podcast, *The Ezra Klein Show*, then part of *Vox Conversations*. The episode is entitled "A Serious Conversation about UFOs" (2019). The full title of Pasulka's book is *Ameri-*

can Cosmic: UFOs, Religion, Technology (Oxford University Press, 2019).

Carl Sagan said "If a human disagrees with you, let him live. In a hundred billion galaxies, you will not find another" in his highly recommended series *Cosmos: A Personal Voyage* (Carl Sagan Productions and KCET, broadcast by PBS in 1980). There is now an updated version of the program starring the astrophysicist Neil deGrasse Tyson (*Cosmos: A Spacetime Odyssey*, broadcast by Fox and National Geographic Channel, 2014).

And don't miss the short documentary *John Was Trying to Contact Aliens* (Matthew Killip, 2020). Still available (at this writing) on Netflix.

18. NIGHT WATCH

John Berger's quote comes from his book *And Our Faces, My Heart, Brief as Photos* (Pantheon Books, 1984).

The glowworms we saw in the Vliegenbos were *Lampyris noctiluca*, the common glowworm of Europe. It was unusual to spot them in October, as their season usually goes until September, when they go into hibernation. Glowworms are in fact not worms at all, but beetles that have always gone by other names. Consult your favorite nature website for more information about this creature.

Nightwatch Noord is starting to get off the ground. Najat Kaddour has, to my delight, joined the project and by the time this book has gone to print, the night walks will have taken place. Interested in joining a night walk? Sign up on my website (marjolijnvanheemstra.nl) or follow me on Instagram, where I share all the news, events, and announcements concerning Night Watch.

From March 2018 until May 2020 I regularly published items on space and space travel on the platform of *De Correspondent*. Some of the information in this book can also be found, in one form or another, in those articles. All my writing for *De Correspondent* can be accessed via decorrespondent.nl/marjolijnvanheemstra.

Index